TWENTY FIRST CENTURY
SCINCE

GCSE Physics

Nuffield
Curriculum Centre

OCR
RECOGNISING ACHIEVEMENT

THE UNIVERSITY *of York*

OXFORD

The exercises in this Workbook cover the OCR requirements for each module. If you do them during the course, your completed Workbook will help you revise for exams.

Project Directors
Jenifer Burden
John Holman
Andrew Hunt
Robin Millar

Project Officers
Peter Campbell
Angela Hall
John Lazonby
Peter Nicolson

Course Editors
Jenifer Burden
Peter Campbell
Andrew Hunt
Robin Millar

Authors
Peter Campbell
Charles Tracey

Contents

P1   The Earth in the Universe   3
P2   Radiation and life   18
P3   Radioactive materials   33
P4   Explaining motion   48
P5   Electric circuits   63
P6   The wave model of radiation   78
P7   Observing the Universe   94

WORKBOOK

OXFORD
UNIVERSITY PRESS

Great Clarendon Street, Oxford OX2 6DP

Oxford University Press is a department of the University of Oxford.
It furthers the University's objective of excellence in research, scholarship,
and education by publishing worldwide in

Oxford   New York

Auckland   Cape Town   Dar es Salaam   Hong Kong   Karachi
Kuala Lumpur   Madrid   Melbourne   Mexico City   Nairobi
New Delhi   Shanghai   Taipei   Toronto

With offices in

Argentina   Austria   Brazil   Chile   Czech Republic   France   Greece
Guatemala   Hungary   Italy   Japan   Poland   Portugal   Singapore
South Korea   Switzerland   Thailand   Turkey   Ukraine   Vietnam

© University of York on behalf of UYSEG and the Nuffield Foundation 2006

British Library Cataloguing in Publication Data

Data available

ISBN: 978-0-19-915053-3

10 9 8 7 6 5 4 3 2

Printed in Spain by Unigraf

Illustrations by IFA Design, Plymouth, UK

These resources have been developed to support teachers and students undertaking a new OCR suite of GCSE Science specifications, Twenty First Century Science.

Many people from schools, colleges, universities, industry, and the professions have contributed to the production of these resources. The feedback from over 75 Pilot Centres was invaluable. It led to significant changes to the course specifications, and to the supporting resources for teaching and learning.

The University of York Science Education Group (UYSEG) and Nuffield Curriculum Centre worked in partnership with an OCR team led by Mary Whitehouse, Elizabeth Herbert and Emily Clare to create the specifications, which have their origins in the Beyond 2000 report (Millar & Osborne, 1998) and subsequent Key Stage 4 development work undertaken by UYSEG and the Nuffield Curriculum Centre for QCA. Bryan Milner and Michael Reiss also contributed to this work, which is reported in: 21st Century Science GCSE Pilot Development: Final Report (UYSEG, March 2002).

Sponsors
The development of Twenty First Century Science was made possible by generous support from:
• The Nuffield Foundation
• The Salters' Institute
• The Wellcome Trust

wellcometrust

THE SALTERS' INSTITUTE

1 The big picture – time and Space

a Here is a diagram of the Earth.
Add these labels:

core – solid   mantle
core – liquid   crust

b The Universe has parts that sit one inside the other.
Place the following objects in order, from smallest to largest structures.

Universe   solar system   planet   galaxy

smallest   largest

Planet	Solar System	Galaxy	Universe

c Put the following in order of age, with the oldest first:

Sun   Earth   Universe   _Universe, Sun, Earth_

d Put the following in order of size (diameter), with the largest first:

Sun   Milky Way   Earth   _Earth   Sun   Milky Way_

e Write a few sentences to describe how the Solar System was formed.

There was a group of clustered stars which collided causing a supernova called the "Big Bang". Rocks formed into planets and started orbiting the sun

f Scientists know the answers to some questions but not others.
Put the letter for each question into the correct column of the table:

A Is there life in the Universe anywhere other than on Earth?

B Did any people live during the time of the dinosaurs?

C How long did it take for multi-celled organisms to develop from the earliest single-celled organisms?

D What is the Sun made from?

E Where and how does the Earth move through Space?

Known	Unknown

2   James Hutton and deep time

a Underline or highlight in one colour the statements (or part statements) that are **data**.
Underline or highlight in a different colour the statements that are **explanations**.

Key: ☐ data   ☐ explanations.

E ⮞ Without some way of building new mountains, erosion would wear the continents flat.

D ⮞ Rivers carry sediment to the oceans, where it settles at the bottom as sand and silt.

E ⮞ Sediments are compressed and cemented to form sedimentary rocks. In some places, layers of
sedimentary rocks are tilted or folded.

b Write a caption for the diagram below. Your caption should describe what the diagram shows.

A beach location with fengrained drawings of fossils fuds.

c Rocks provide evidence for long-term changes in the Earth.
For example, some rocks in Britain contain fossils of creatures that once lived in tropical areas.
This is evidence of continental drift.

Describe two other ways that rocks provide evidence of changes in the Earth.

1   The different types of rock

2   Erosion, weathering.

3 Alfred Wegener and continental drift

a Early in the twentieth century, Alfred Wegener suggested the idea of continental drift to explain mountain building. Complete the continent labels on the diagram.

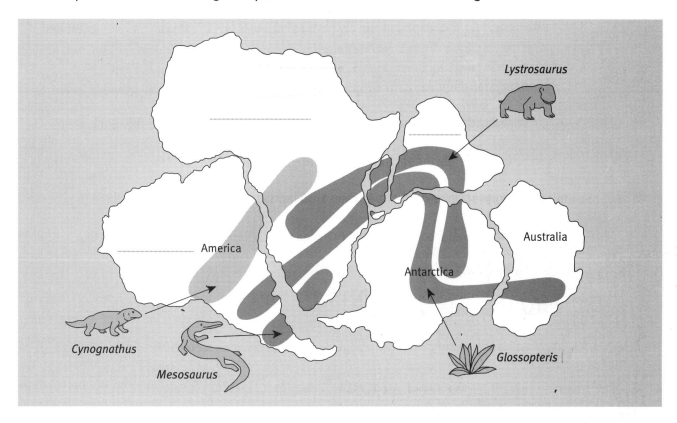

b Complete these sentences.

Wegener first presented his idea of continental drift to other scientists at a conference in 1912. His book,

titled _The Origin of Continents and Oceans._, described the idea and evidence for it. This enabled

other scientists to evaluate his work, a process called _Continental drift_.

c How did Wegener explain mountain building?

Number the following statements to put them in the correct order. The first one has been done for you.

| 1 | Erosion gradually wears done mountains. |

| | A drifting continent ploughs across the ocean floor. |

| | The wrinkled edge is a chain of mountains. |

| | Without some way of building new mountains, the continents would be worn until they were completely flat. |

| | Good examples are the Rocky mountains of North America and the Andes mountains in South America. |

| | Friction with the ocean floor produces a wrinkle at the leading edge of the moving continent. |

d Wegener's idea of continental drift started a scientific debate that lasted several decades.

Use words from the list to complete the following sentences, which summarize the debate.

big	measure	explanation	simpler	data	geologist

➜ The _____ itself was accepted as correct by other scientists.

➜ Continental drift was a possible _explanation_ of the how continents could move,

by ploughing across the ocean floor.

➜ Most scientists disagreed with Wegener's explanation because:

  ● the movement of continents was too small to _measure_

  ● there were _simpler_ explanations of the same data

  ● he was meteorologist and not a _geologist_

  ● the idea of moving continents was too _big_ an idea from limited evidence.

e Some of the endings for the following statement are correct and some are wrong.
Put a tick ✔ next to any that are correct.

A good scientific explanation is one that

➜ comes with a diagram ☑

➜ accounts for all observations ☒

➜ is not evaluated by other scientists ☒

➜ links things that were previously thought unrelated ☑

➜ is boldly and clearly stated ☑

➜ leads to predictions that are later confirmed ☑

4 Fred Vine and seafloor spreading

a The diagram shows an oceanic ridge.

Use the diagram to help you describe the part played by the solid mantle in seafloor spreading.

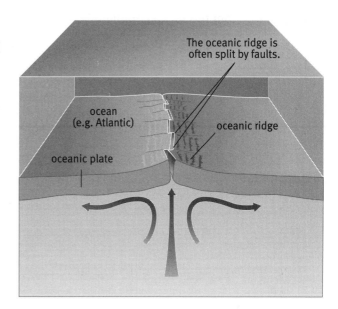

The oceanic ridge is often split by faults.

ocean
(e.g. Atlantic)

oceanic ridge

oceanic plate

....

....

....

....

....

....

b The direction of the Earth's magnetic field changes from time to time. When magma solidifies, the direction of the magnetic field is preserved in the rocks.

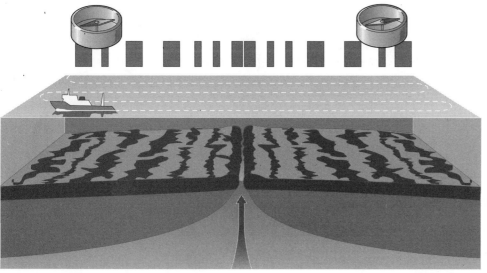

i Describe the pattern of magnetic field differences on either side of the mid-oceanic ridge.

....

....

ii Explain how this pattern provides evidence for seafloor spreading.

....

....

....

7

5 Plate tectonics theory

a Use words from this list to complete the sentences:

convection currents	radioactive decay	tectonic plates	mantle

➔ The Earth's crust is made of a series of plates called _tectonic Plates_ .

➔ The plates move very slowly, carried along by slow _Connvechon Currents_

in the _Maylle_ .

➔ The heat released from _radioactrue_ causes the convection currents.

b

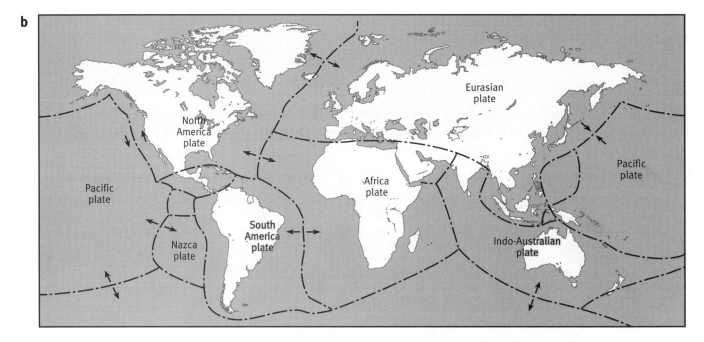

Mark with letters on the map a place where there is: (**M**) a mountain chain, (**R**) an oceanic ridge, (**C**) oceanic crust. Also mark places where (**E**) earthquakes and (**V**) volcanoes are likely to happen.

c Use words from the list to complete the sentences, describing what scientists knew by the mid-1960s.

Although the mantle is _Solidhot_ , it can _flow_ very slowly.

Energy is released when the _spread_ of _radioachve_ elements inside

the Earth _decay_ . So the inside of the Earth stays _Solid_ .

Seafloor _atoms_ happens at _Oceanic_ ridges. This moves

continents apart.

decay

radioactive

solid

flow

atoms

oceanic

hot

spreading

d For each sentence in part **c**, decide whether it is evidence for, or a mechanism for, continental drift. Write **E** (evidence) or **M** (mechanism) next to each sentence.

e Changes in the Earth's surface are very slow. The materials that make up the Earth are constantly being recycled.

Use words from the list to complete the labels on the diagram:

lava flows	sediment	oceanic plate	rock carried into subduction zone

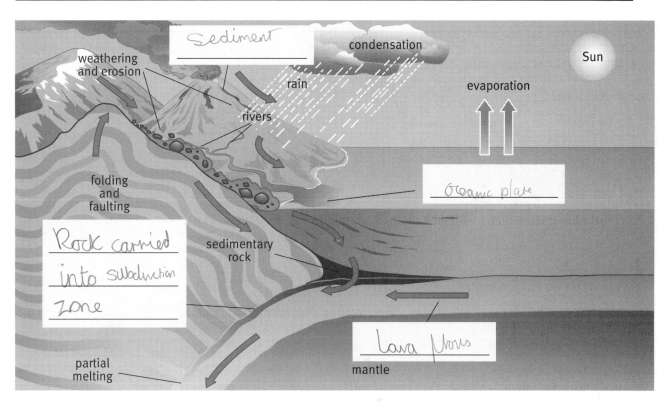

f Describe the part played by the movement of tectonic plates in

? ➔ earthquakes: *The earth shakes violently causing physical destruction*

? ➔ volcanoes: *forms when earthquakes occur leaving a mass of ash and heated molten rock*

➔ mountain building: _____

_____

➔ parts of the rock cycle: _____

_____

6 Geohazards

a Use words from the box to complete the spider diagram:

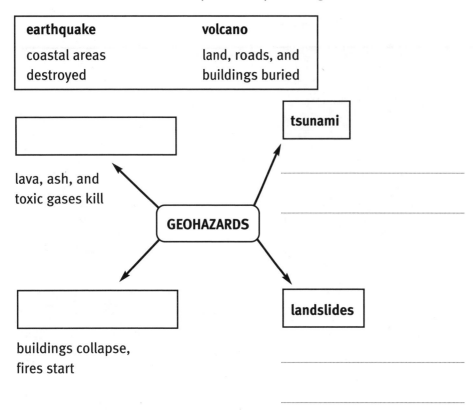

earthquake	volcano
coastal areas destroyed	land, roads, and buildings buried

tsunami

lava, ash, and toxic gases kill

GEOHAZARDS

landslides

buildings collapse, fires start

b Why do scientists search for ways of predicting earthquakes and volcanic eruptions?

It is important because to avoid major disaster on earth and its inhabitants.

c What can public authorities do to reduce the damage caused by volcanoes, earthquakes, or tsunamis?

1   To warn people in advance

2   To clear the disaster area

3   To set up an exclusion zone around the area.

4   Prepare people with the necessary protection gear eg gas mask food and proper shelter

7 Craters – what makes them?

a The Earth is a planet that moves around the Sun. The Earth and the Sun, with other objects, make up the Solar System.

Label on the diagram: the Sun, a planet, the asteroid belt, a comet.

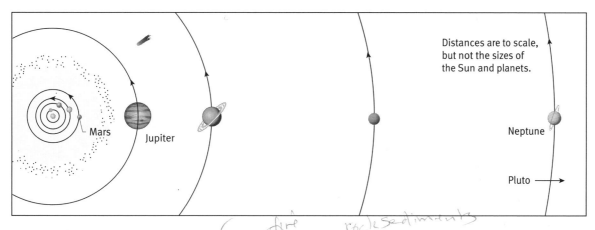

Mars   Jupiter   Neptune   Pluto →

Distances are to scale, but not the sizes of the Sun and planets.

b What is the difference between moons, comets, and asteroids? Describe their relative sizes and motions.

Comets are Stars that are filled with gases eg hydrogen Moons are huge rock stars. Asteroids are pieces of loose rocks that drifts slowly alone in space.

c The Solar System formed from a cloud of dust and gas (mainly hydrogen). About 5000 million years ago, gravity caused the cloud to collapse inwards, creating the Sun and all the objects that orbit it. Explain the evidence and the reasoning that enable scientists to estimate when this happened.

All of this is highly probably. It only gives us an indication. It cannot be proven. The evidence on earth is through the fossilised rocks.

d Complete these sentences.

There is a crater the size of twenty football pitches in ___Nevada___, USA. At first, scientists thought

the crater was made by a ___asteroid___. Daniel Barringer, a mining engineer, found fragments of

___rock/sediments___ in the crater. He concluded that the crater was made by a ___collision___

because they too contain iron. More evidence was found to ___prove___ Barringer's explanation:

- the crater contains quartz dust particles, created by huge ___explosion___
- the layers of rock surrounding the crater are in _____

Today's explanation, which accounts for all the data, is that a ___?___ made the crater.

11

8 What killed off the dinosaurs?

a Explain what a 'mass extinction' means.

A complete eradication on a huge scale

Scientists disagree about the reasons why dinosaurs became extinct. Two theories have been put forward:

↔ asteroid strike

↔ massive eruption

Evidence has been found:

A An iridium layer dated at 65 million years ago has been found in rocks in many parts of the world. Asteroids and planet cores contain iridium in similar concentrations.

B Larger than usual eruptions of lava took place in what is now India between 63 and 68 million years ago.

C Eruptions can release

↔ poisonous gases

↔ dust clouds that block the Sun's heat and light

D Not all large eruptions of lava in the past caused mass extinctions.

E A large impact crater of the right date was discovered in Mexico.

F Extinctions began before the asteroid struck.

b Identify evidence for and against the two theories by writing letters from the list **A–F** in the table.

Theory	For	Against
asteroid strike		
massive eruption		

c Describe how each of these events could cause mass extinctions:

An asteroid strike:

Massive eruptions of lava:

d You have seen the links between asteroid strikes, lava flows, and dinosaur extinction. Why can't you be sure that either, or both, events caused the extinction?

9   What elements are we made of?

a Use a tick ✔ to show the correct ending for this statement.

We know what distant stars and galaxies are made from because . . .

➔ space probes carry robotic laboratories that test them ☐

➔ manned space flights collect samples, returned to Earth for testing ☐

➔ their light reaches Earth ☐

b Ideas have changed about what stars are made from, and where the chemical elements found on Earth come from.

A When astronomers first looked at the spectrum of sunlight they found that it has a pattern of dark lines in it.

B Laboratory experiments produce the same pattern of dark lines as the spectrum of sunlight. This means the same chemical elements are present.

C The Sun is made from chemical elements found on Earth.

D Light from the early Universe shows that the only elements present then were hydrogen, helium, and a little lithium. Dying stars scatter heavier elements through Space.

E Nuclear fusion produces new elements and takes place in stars. Elements found on Earth come from earlier stars. We are made of stardust.

Use these statements to build a story about how science explanations develop.
Put letters next to the statements in the grid below to complete the story. The first one has been done for you.

A scientist makes an observation that needs an explanation.	**A**

A scientist thinks creatively to explain that observation.	

A scientist uses observations to support an idea.	

Scientists develop the explanation further and predict other outcomes.	

Later new observations are made by scientists that support the same explanation.	

The explanation becomes widely accepted.

10 Are we alone?

a Light travels at .... km/s. Yet because of the huge distances, its time of travel through
Space can be large.

Complete the table showing how long ago light left each object.

Object	When light reaching Earth left it
Sun	
Proxima Centauri	
Arcturus	

b What general pattern does this show about distances and the observed age of the objects that
astronomers study?

....

....

c Describe what a light-year is.

....

d Two methods of measuring the distances to stars and galaxies are

1 ....

2 ....

e Why is there uncertainty in the measured distances to stars and galaxies?

....

f Give an argument to convince someone that life elsewhere in the Universe is likely.
(An argument will involve both evidence and reasoning.)

....

....

g Give an argument to convince someone that life elsewhere in the Universe is unlikely.

....

....

11 The great debate – Shapley v Curtis

a Light pollution is a problem for anyone wanting to see more than just the brightest stars in the night sky.

Draw lines to complete the statements that explain why.

Streetlights and car parks . . .	. . . make the night sky glow brightly.
Upward-travelling light scatters off particles in the atmosphere to . . .	. . . become impossible to see.
Faint, deep sky objects such as galaxies and nebula . . .	. . . direct light upwards into the night sky.

b The 'great debate' of 1920 concerned nebulae. At the time, about 15 000 nebulae had been observed.

Some astronomers thought nebulae were located _____

Others thought nebulae were located _____

c Tick ✔ the statement that best completes this conclusion.

The outcome of the debate was overturned by Hubble's observations a few years later because . . .

↪ some of the observations that Curtis presented were incorrect ☐

↪ Hubble was a better scientist than Curtis or Shapley ☐

↪ Hubble's telescope enabled him to make a better measurement of the distance to Andromeda ☐

d Complete the labels on the chart:

the Universe

↑

billions of _____

(many times their own diameter apart)

↑

our own galaxy, the _____

(100 000 _____ in diameter)

↑

the Solar System

(The _____ is just one of billions of _____ in the Milky Way)

12 How did the Universe begin?

a Complete the sentences and provide additional information to make your own notes about how the Universe began.

➜ Dark lines in the spectra from distant galaxies are shifted towards the _____ end of the spectrum. This shows that they are moving away from us.

➜ The more distant a galaxy, the _____ it is moving away.

➜ If galaxies are moving further and further apart, the Universe must be _____ .

➜ Write a caption for the diagram above, to explain the previous statement.

_____

_____

_____

➜ One _____ for this expansion is that Universe began with a huge explosion about _____ years ago. This is called the _____ _____ theory.

b Three independent lines of evidence for the big bang theory are

1 ....

2 ....

3 ....

One of these is an observation that was predicted from the theory before it was observed. Mark it with an **O**.

Another was observed first and then explained by big bang theory. Mark this **F**.

c Describe the process of peer review.

....

....

....

d Peer review makes scientific data and observations more reliable. Explain how.

....

....

....

e The ultimate fate of the Universe is difficult to predict. Explain why.

....

....

....

f Stephen Hawking is a scientist famous for his theories about black holes and the Universe. In what way does he provide a good example of the role of human imagination in developing scientific explanations?

....

....

....

1 Taking chances with the Sun

a The Sun's radiation is all around you. Three types of the Sun's radiation are shown below.

i Draw a line to match each type of radiation with its description on the right. One has been done for you.

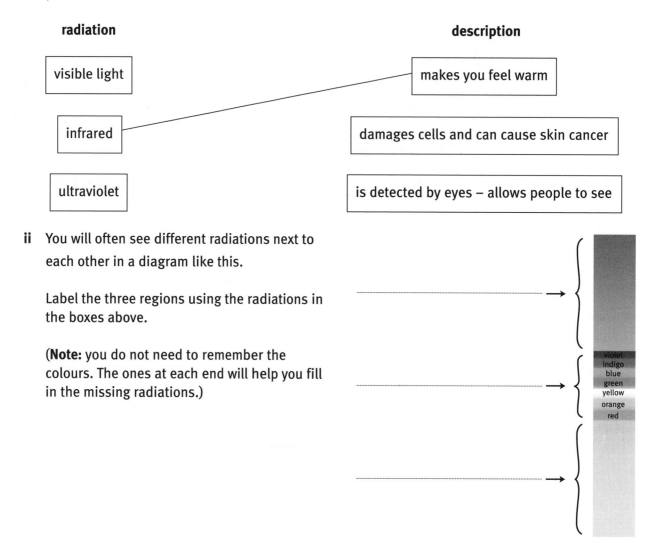

radiation

description

| visible light | | makes you feel warm |

| infrared |

| ultraviolet | | damages cells and can cause skin cancer |

| | is detected by eyes – allows people to see |

ii You will often see different radiations next to each other in a diagram like this.

Label the three regions using the radiations in the boxes above.

(**Note:** you do not need to remember the colours. The ones at each end will help you fill in the missing radiations.)

violet
indigo
blue
green
yellow
orange
red

iii When radiation hits something, the material can either **reflect** or **absorb** or **transmit** the radiation.

Look at the diagrams below. Decide what happens to visible light when it hits these objects.

Label the diagrams using the words in the list.

| transmit | reflect | absorb |

You can use one or more words on each diagram.

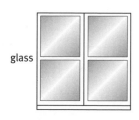

glass

paper

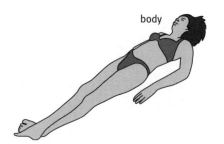

body

b There are benefits and risks of being exposed to sunlight.

Write down two benefits and two risks in the table. One has already been done for you.

Benefits	Risks
Makes me feel good	

c Look at these statements.

A A survey in 2006 found that children who are polite do better in exams.

B Traffic data shows that you are more likely to have a car accident if you have been drinking alcohol.

i In each statement above:

↦ underline the outcome.

↦ draw a ring around the possible factor.

ii Complete these sentences by drawing a ring around the correct **bold** words.

↦ In statements A and B, there is a **correlation / approximation** between a factor and an outcome.

↦ In statement B, drinking alcohol is the **cause / effect** of an increase in car accidents.

↦ In statement **A / B**, it is not necessarily the case that the factor causes the outcome.

d You and a friend are planning to spend a summer's day on the beach. You put on sunscreen.

i Complete these sentences. Fill in the missing words. Put a ring round the correct **bold** words.

SPF stands for S_____ P_____ F_____. High SPF sunscreen absorbs much of the **ultraviolet / infrared** radiation from the Sun. Using a **high / low** SPF sunscreen will **increase / decrease** the amount of time it takes for radiation to cause skin redness. This will **increase / decrease** a person's risk of getting skin cancer.

ii Write down one other way to protect yourself when you are in sunlight.

2 Sunlight and life

a i   Use these words to complete the sentences in the boxes on the left.
You can use each word once, more than once, or not at all.

| atmosphere | ozone layer | plant | transmits | absorbs |

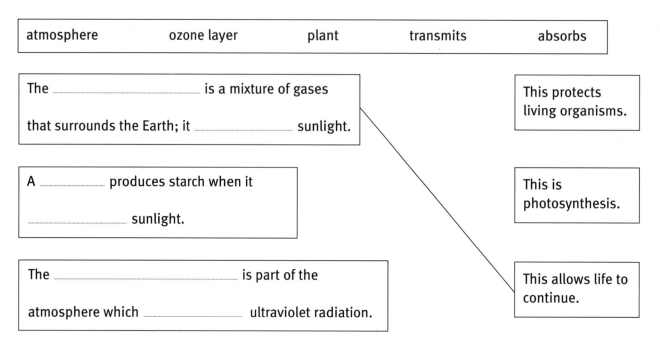

The _____ is a mixture of gases that surrounds the Earth; it _____ sunlight.

This protects living organisms.

A _____ produces starch when it _____ sunlight.

This is photosynthesis.

The _____ is part of the atmosphere which _____ ultraviolet radiation.

This allows life to continue.

ii   Join up the boxes to makes three statements about radiation and the Earth. The first one has been done for you.

b The ozone layer is important to life on Earth. Complete the diagram to explain why.

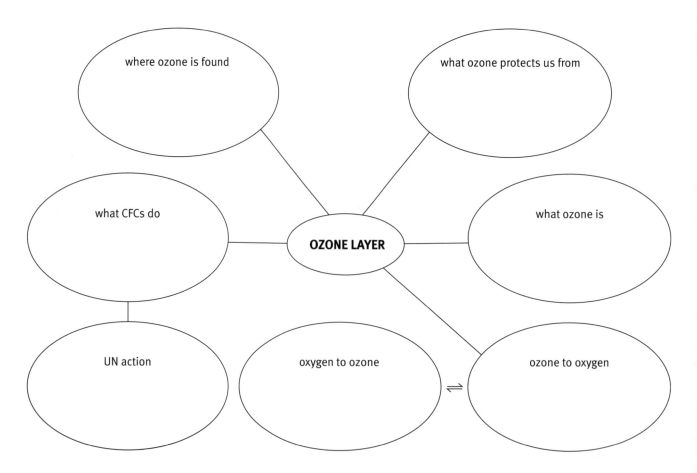

where ozone is found

what ozone protects us from

what CFCs do

what ozone is

OZONE LAYER

UN action

oxygen to ozone

ozone to oxygen

3   Radiation journeys

a The picture shows a rescue worker using a special camera to find survivors. The diagram on the right is a model for describing radiation.

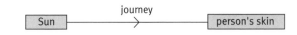

i   Use the model to help you answer the following.

⇥ What type of **radiation** is the rescuer searching for? _____

⇥ What is the possible **source** of this radiation? _____

⇥ What is the **detector** for the radiation? _____

ii   Some parts of the journey will absorb the radiation and some parts will transmit it. Give one example of a material that

⇥ absorbs the radiation: _____

⇥ transmits the radiation: _____

b Look at how the model is applied to ultraviolet radiation coming from the Sun to a person on a beach.

Complete the sentences. Draw a (ring) round the correct **bold** words.

The **Sun** / **skin** is the source which **emits** / **absorbs** ultraviolet radiation. This travels through Space and the Earth's **atmosphere** / **oceans** before it is **absorbed** / **emitted** by the **Sun** / **skin**, which is the detector.

c Look at the statements below. They describe the process called **selective filtering**. But they are mixed up. Put a number next to each statement to show the correct order. The first one has been done for you.

☐ For example, the ozone layer absorbs some ultraviolet radiation

☐ This is an example of filtering. It is called selective filtering

☐ because only selected frequencies are absorbed.

1 Some materials absorb only certain frequencies of radiation.

☐ but transmits visible light.

4 Absorbing electromagnetic radiation

a Some radiations can damage cells in one of two ways. Look at the paragraph below and draw a ring round the correct **bold** words.

Things **heat up** / **cool down** when they absorb radiation. Greater amounts of radiation cause **more** / **less** heating. This heating may damage living things. **High** / **Low** energy radiation can also cause damage to cells directly – not just by heating them up. It damages the cells because it causes **ionization** / **emission**. The radiation breaks up **molecules** / **vessels** in the cell, producing charged fragments called **ions** / **currents**. These can take part in **opposite** / **chemical** reactions in the cell and disrupt its behaviour.

b Look at the radiations in part **d** below. They are all part of the same family of radiations. This family has a special name.

Draw a ring around the two words that make up its name (choose one word from each list).

List A: magnetic   light   electromagnetic   inductive   multicoloured

List B: induction   family   relations   rainbow   spectrum

c Radiation is energy given out by a source. The energy is carried in packets.

What is the name of a packet of electromagnetic radiation? ....

d The amount of energy in a photon is different for each type of radiation.

   i Place these types of radiation in order in the diagram.

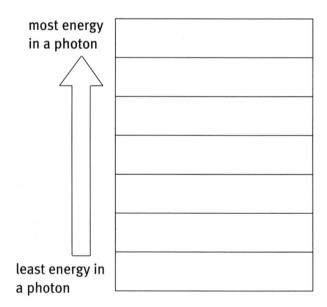

radio waves

X-rays

microwaves

gamma rays

infrared

visible light

ultraviolet

most energy in a photon

least energy in a photon

   ii Shade the boxes that represent ionizing radiation.

   iii Put a letter 'H' next to the boxes of radiations which can cause damage by heating.

5 Microwave safety

Microwave ovens use the heating properties of microwave radiation to cook food.

a Different materials have a different effect on microwaves. Look at the materials (on the left).

➲ Draw lines to match each material with its effect on microwaves (in the middle). One has been done.

➲ Draw three more lines to match each effect with the role it plays in a microwave oven (on the right).

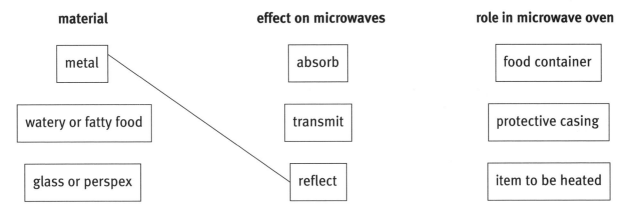

material	effect on microwaves	role in microwave oven
metal	absorb	food container
watery or fatty food	transmit	protective casing
glass or perspex	reflect	item to be heated

b Neil buys a new microwave oven. It has a power rating of 100 watts. This means it transfers 100 J of energy per second. His old microwave oven was rated at 60 watts.

i Which microwave oven produces radiation with a higher intensity? ....

ii Will his new oven cook a potato in more or less time? ....

iii His old oven took 10 minutes to bake a small potato. How long will his new oven take ....

c The electric toaster was invented about a hundred years before the microwave oven. Previously, bread was toasted on a fork held up to an open fire. Like all new devices, the electric toaster is not completely safe.

i Use the see-saw diagram to list some risks and benefits of electric toasters.

risks   benefits

(**Note:** you will not need to remember these; it is the *method* of weighing risks and benefits that you need to understand.)

ii Do the benefits outweigh the risks or the other way around? ....

iii How does the see-saw diagram help you evaluate whether it is worth taking a risk?

....

iv What has happened to the perceived risk over time? ....

6 Using mobile phones

a Mobile phones transmit and receive phone calls and text messages using electromagnetic radiation.

   i Which part of the electromagnetic spectrum do they use? _____

   ii Some people are concerned about the effects of this radiation. Their strategy is to use a microphone and earpiece on the end of a long piece of wire rather than holding the handset next to their ear. What effect does this strategy have on the intensity of the radiation by their head?

   _____

   iii Suggest why they do this.

   _____

   iv Draw a ring round the correct **bold** words to make a phrase that describes their strategy.

   Better **safe** / **quiet** than **noisy** / **sorry**.

   v It's not always easy to decide how much of a risk something is. When you are not sure, it is best to use this approach. Look at the names below. Put a ring round the one that is given to this approach.

   | ALARA | Cole's Law | Precautionary Principle | Murphy's Law | Pollution control |

b Explain the scientist's statement – why is nothing ever completely safe?

   _____

   _____

   _____

We're collecting more evidence about mobile phone safety. But whatever we find out, we'll never be able to say that they're **completely** safe.

c People often estimate risks for things they do. Their judgement is called the **perceived risk**. But this may not be the **actual risk**.

Write a 'P' by any statement(s) where the **perceived** risk is bigger than the actual risk.
Write an 'A' by any statement(s) where the **actual** risk is bigger than the perceived risk.

I love downhill skiing. The steeper the slope, the better! It's perfectly safe if you know what you're doing.

I work on a building site. We're supposed to wear hard hats but I don't know of any accidents where a hard hat helped. If the boss isn't around, I take mine off. It's cooler.

Have you heard about BARS, a new infectious disease? Until scientists work out how to stop it spreading, I'm not using lifts or going on tube trains full of people who may be infected – however inconvenient!

I don't like flying. Too risky! When I go on holiday I always drive, even though the journey may take days.

7 Health studies

Frances claims that she can read people's minds. She asks them to think of a number.
She says that she can tell whether they are thinking of an even number or an odd number.

a She tried this out on eight of her friends. She repeated it four times on each one. These are her results:

Friend	A	B	C	D	E	F	G	H
Number of correct readings (out of 4)	1	4	3	2	1	3	3	2

i For how many children did Frances get: 100% correct ....   75% correct ....

Georgia says she does not believe Frances. She says it was just chance. She tries flicking a coin four times and counts how many heads she gets. She repeats this eight times. These are her results:

Number of heads out of four flicks:	3	1	3	2	4	1	0	3

ii For how many tests did Georgia get: 100% heads ....   75% heads ....

iii Explain her point about the role of chance.

....

b Frances says that it is only people with dark hair who are telepathic (able to have their minds read). They give out 'tele-waves'. Friends B, C, F, and G have dark hair. The others have blond or red hair.

i Look again at the table in part **a**. Draw a (ring) round the friends who have dark hair.

ii Is there a correlation between having dark hair and being telepathic? ....

iii Is there enough evidence to say that dark hair makes people telepathic? Explain your reasoning.

....

iv Frances chose the factor of dark hair after the experiment. Suggest two other possible factors.

1 ....   **2** ....

v What is the mechanism that Frances has suggested? ....

c Georgia asks Frances to repeat the experiment but to do 20 tests on each friend. These are the results:

Number of correct readings out of 20:	14	9	11	8	7	13	10	6

i For how many children did Frances get: 100% correct ....   75% or more correct ....

ii Complete this paragraph by drawing a (ring) round the correct bold words.

More readings make the results **more / less** reliable. This is because the likelihood of getting a high score by chance has **increased / decreased**.

8 X-ray safety

X-rays are an ionizing radiation.

a Ionizing radiation causes damage to cells because it makes molecules more likely to react. These reactions can make toxic products.

Match the start of each statement to its best correct ending. One has been done for you.

When energy carried by microwaves is absorbed . . .	. . . it can cause chemical changes.
Radiation is called 'ionizing' if . . .	. . . it may start a cancer tumour.
If ionizing radiation damages a cell's DNA . . .	. . . it kills the cells.
When large amounts of ionizing radiation strike living cells . . .	. . . it causes a heating effect.

b Doctors know that having an X-ray increases the chance of getting a cancer in later life. The increased chance is very small – less than 1 in 1 000 000.

Look at the possible uses of X-rays below. Complete the table by answering the questions on the left.

	Using X-rays to check if a child's shoe fits property	Using X-rays to check a broken bone
What are the benefits? (Give at least one.)		
Do the benefit(s) outweigh the risk?		
Would you take the risk? Give your reasons.		

c The ALARA principle means finding a balance between reducing the risk of some activity and the cost of doing this.

i Complete the table to show what ALARA stands for.

ii Write down one example of where you could use the ALARA principle.

A	
L	
A	
R	
A	

9　The greenhouse effect

The Earth's atmosphere heats up when it absorbs radiation from the Earth. This is called the Earth's **greenhouse effect**.

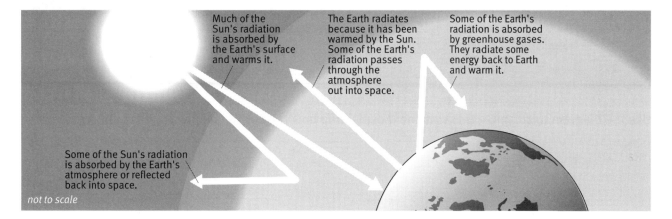

Much of the Sun's radiation is absorbed by the Earth's surface and warms it.

The Earth radiates because it has been warmed by the Sun. Some of the Earth's radiation passes through the atmosphere out into space.

Some of the Earth's radiation is absorbed by greenhouse gases. They radiate some energy back to Earth and warm it.

Some of the Sun's radiation is absorbed by the Earth's atmosphere or reflected back into space.

not to scale

a Look at the three boxed objects on the left below.

i Draw lines to match each object with its description in the middle. The first one has been done for you.

ii Draw three more lines to match each description with how it emits or transmits radiation.

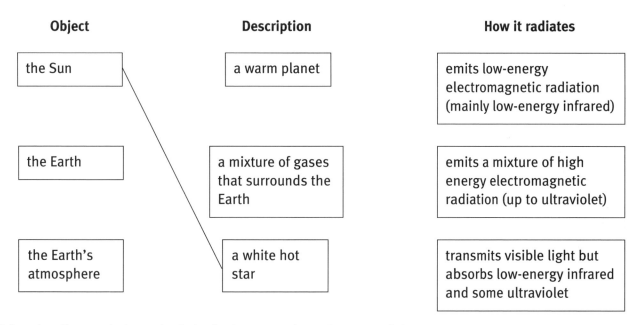

Object	Description	How it radiates
the Sun	a warm planet	emits low-energy electromagnetic radiation (mainly low-energy infrared)
the Earth	a mixture of gases that surrounds the Earth	emits a mixture of high energy electromagnetic radiation (up to ultraviolet)
the Earth's atmosphere	a white hot star	transmits visible light but absorbs low-energy infrared and some ultraviolet

b Using the diagram below, shade in the boxes to show the parts of the spectrum

⤵ present in the Sun's radiation
⤵ that the Earth radiates

Sun's radiation

Earth radiates

← ultraviolet　　　　　visible light　　　　　infrared →

electromagnetic spectrum

27

The Earth's greenhouse effect has prevented the Earth from getting too cold. The picture shows the balance between the energy it absorbs and the energy it radiates.

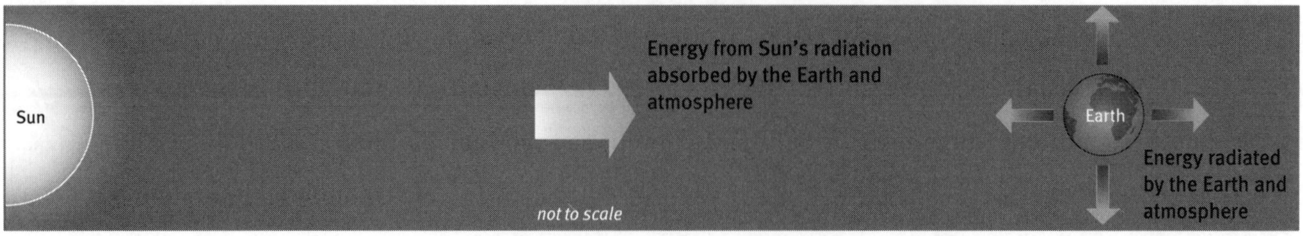

not to scale

Energy from Sun's radiation absorbed by the Earth and atmosphere

Energy radiated by the Earth and atmosphere

c Use the diagram to complete this statement explaining how the Earth's temperature is kept stable.

....

....   **=**   ....

....

d Tiny amounts of some gases in the atmosphere cause the greenhouse effect. They absorb radiation from the Earth and heat up.

i Tick ✓ the greenhouse gases in the list below.

☐ methane ☐ oxygen ☐ carbon dioxide ☐ water vapour ☐ nitrogen

ii Complete these sentences by drawing a ring round the correct **bold** words.

Human activity is **increasing** / **decreasing** the amount of some greenhouse gases.

This causes an **increase** / **decrease** in the greenhouse effect which will **raise** / **lower** the average

temperature of the Earth's atmosphere. This will **change** / **stabilize** climates around the Earth.

e Without the greenhouse effect, the Earth would be too cold for life.
But if the greenhouse effect were too great, it would be too hot.

The table below compares Earth **with** and **without** its atmosphere, and with an increased greenhouse effect. Complete the table. Put a ring round the correct bold words. Use these words to fill in the gaps.

life cannot exist   life can exist   −18   15   17   the climate will change

Earth with an atmosphere	Earth without an atmosphere	Earth with increased greenhouse effect
Average temperature .... °C	Average temperature .... °C	Average temperature .... °C
Water is **solid** / **liquid** / **vapour**	Water is **solid** / **liquid** / **vapour**	This means
This means	This means	

10 The carbon cycle

The diagram shows the carbon cycle. The boxes are stores of carbon in the environment. The arrows show ways that carbon moves between them.

a Complete the diagram using these words.

oceans	atmosphere	animals	plants	fossil fuels
respiration	decomposing	eating	photosynthesis	burning fuels

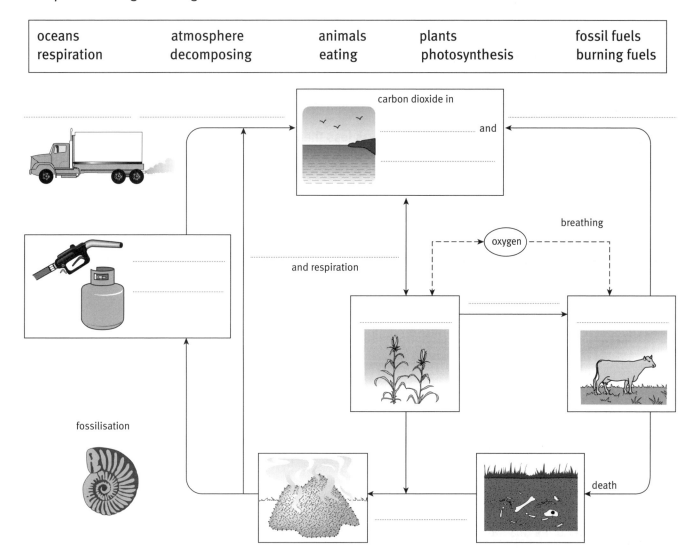

➜ Draw a blue ring around the one process that removes carbon dioxide from the atmosphere.

➜ Draw a green ring around the three processes that have been returning carbon dioxide to the atmosphere for thousands of years.

➜ Draw a red ring around a process that has been returning carbon dioxide to the atmosphere for the last two hundred years.

b Complete this sentence. Draw a ring round the correct **bold** words.

Decomposers / **cows** help to recycle carbon by **releasing** / **absorbing** carbon dioxide from the

carbohydrates / **oxygen** in dead and decaying animals and plants.

The amount of carbon dioxide in the atmosphere was stable for thousands of years.
About 200 years ago, it began to rise.

c Look at the word equation below. It explains why the amount of carbon dioxide was constant until about 200 years ago. Complete the word equation by

➜ writing in one process on the left and three on the right

➜ drawing a (ring) round the correct **bold phrase** in the middle

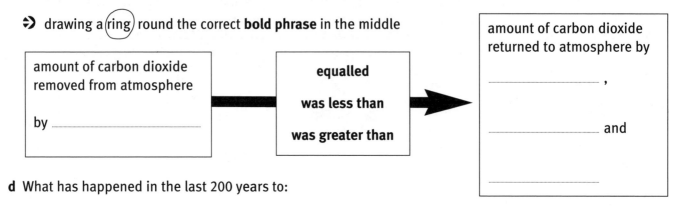

amount of carbon dioxide removed from atmosphere

by ....

equalled

was less than

was greater than

amount of carbon dioxide returned to atmosphere by

.... ,

.... and

....

d What has happened in the last 200 years to:

➜ increase the amount of carbon dioxide returned to the atmosphere?

....

➜ decrease the amount of carbon dioxide removed from the atmosphere?

....

e As a result of these two changes, the phrase in the middle box above may have changed.

➜ Draw a ring round the new phrase to go in the middle box.

equalled	is less than	is greater than

➜ What effect does this have on the amount of carbon dioxide in the atmosphere?

....

f Complete these sentences. Fill in the missing words. Put a (ring) round the correct **bold** words.
Use the carbon cycle on page 44 to help you.

Carbon dioxide is one greenhouse gas. There are very **small / large** amounts of carbon dioxide in the

atmosphere. Green plants use .... from the Sun's radiation to make their food. They do this

by the process of .... This **uses up / releases** carbon dioxide.

g Complete the word equation for photosynthesis:

....

.... + ... ⟶ .... + ...

11 Changing climates? Possible consequences

a Look at the statements below. Label each one with **climate** or **weather** to show the difference it is describing.

↪ The 6th March 2006 was sunny in London but it rained the next day. _____

↪ The UK is quite temperate, whereas southern Spain is hot and dry. _____

↪ Farmers are planting summer crops 3 weeks earlier than they did 50 years ago. _____

↪ The annual rainfall in Edinburgh was less in 2005 than it was in 2006. _____

b Look at the observed effects on the right.

↪ Draw a (ring) round each effect that is an indicator of recent global temperature rises.

↪ Underline each effect that provides evidence of a link between temperatures and carbon dioxide in the atmosphere.

melting glaciers
data from ice cores
growth rings in trees
melting polar ice-caps
annual temperature records
rising sea levels

c We can predict how global warming will affect the planet. It is much more difficult to predict the effects of climate change in any particular region.

Look at the statements below. Draw a line to link each predicted effect of climate change (on the left) with a description of how it may affect humans (on the right). The first one has been done for you.

Predicted effect of climate change

rising sea levels

extreme weather including heavy rains and flooding

disruption to the warm air and ocean currents from the tropics

average temperature rises

How this may affect humans

temperatures drop in UK and northern Europe

low-lying cities flooded and farmland covered

mosquitoes carrying malaria move to currently temperate areas

damage to property and rising insurance premiums

12 Changing climates – time for action?

a i Draw a line to link each statement with one of the boxes on the right.
The first one has been done for you.

> The Earth is warming up – an effect called global warming.

> Future emissions of carbon dioxide will raise global temperatures by about 3 °C in the next 60 years.

> The UK will have wetter winters and dryer summers.

> Global warming is a harmless variation in temperature.

> Human activities contribute to global warming.

> If carbon dioxide concentration rises above 500 ppm, climate change will become irreversible.

> Weather conditions in many parts of the world have become more extreme in the last 10 years.

> To stabilize climates, carbon emissions would need to be reduced by 70% globally.

> Scientists agree about this.

> Scientists do not all agree about this.

ii Some of the statments on the left are predictions made by computer models. Put a tick ✔ next to these statements.

b People and governments can take action to try to avert the potential effects of climate change.
List two actions that can be taken by

➔ individual people ....

➔ governments ....

c The effects of climate change are not evident until decades after the actions that cause it.
Describe how this affects

➔ the time it takes to establish and agree a theory ....

....

➔ how easy it will be to test the effectiveness of any remedies ....

....

1 Energy patterns

a Some homes have a gas boiler to heat their water. Others use an electric immersion heater. The diagrams below show the total efficiency of using each of these methods.

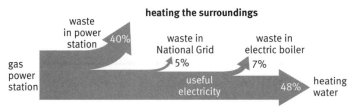

Heating water using an electric immersion heater

Heating water using a gas boiler

i Put a tick ✓ next to the method that is more efficient.

ii What percentage of energy from the primary source is wasted when using

➔ a gas boiler _20_ %   ➔ an electric immersion heater _52_ %

iii When someone uses an electric immersion heater, where is most of the energy wasted? _most of_ _the energy is initially wasted in the power station_

b Gas is a primary energy source.

i Which one of these is *not* a primary energy source? Draw a (ring) round your choice.

| coal | oil | nuclear | (electricity) |

ii Explain why your choice is called a secondary energy source. _I think this is a secondary_ _energy source because there are processes to go through_ _before electricity is made._

uses a primary source.

c Look at the statements below. Tick ✓ those that are advantages of using electricity. The first one has been done for you.

- [✓] It is convenient – comes on at the flick of a switch.
- [ ] It is more efficient than using a primary source.
- [✓] It produces no pollution where it is used.
- [ ] It reduces the amount of carbon dioxide in the atmosphere.
- [✓] It can easily be transmitted over long distances.
- [ ] It can be used by a wide variety of devices.

2/3

d Complete these sentences. Draw a (ring) round the correct **bold** words.

The UK's demand for electricity is **increasing** / **decreasing**. This means that energy companies have

to be able to **generate** / **use** more electricity – especially at peak times such as 6 pm on a

summer's / **winter's** day. If they cannot meet the demand, then there will be **backups** / **blackouts**. So these

companies are building new power stations. Most of these burn **fossil** / **nuclear** fuels which release

oxygen / **carbon dioxide**. This is a **'greenhouse'** / **'ozone'** gas and contributes to changes in

climate / **UV-levels** around the world.

15/17

2 Radiation all around

You are continually exposed to background radiation. Radiation dose is a measure of the possible harm to your body's cells caused by radiation.

Josy calculated the radiation dose she received last year.
Her estimates (in millisieverts, mSv) are shown in the table below.

Source of radiation	Josy's dose (mSv)	Natural or artificial?	Occupations with increased dose
radon in air where she lives	0.700		
from rocks and buildings	0.030		
cosmic rays where she lives	0.230		
cosmic rays from air travel	0.020		
from food and drink	0.300		
from medical treatments	0.040		
from nuclear industry and fallout	0.017		
Total	**1.337**		

a Write either **N** or **A** in the third column to show which sources are **N**atural and which are **A**rtificial.

b What is Josy's total dose from   ➔ natural sources? .... mSv   ➔ artificial sources? .... mSv

c What effect do the artificial sources have on the risk to Josy from radiation? Draw a ring round your choice.

a large increase	a small increase	no effect	a small decrease

d The national average radiation dose is 2.600 mSv for one year. Compared with the national average, suggest the size of the risk to Josy from radiation. Draw a ring round your choice.

more than the national average	less than the national average	much less than the national average

Large radiation doses can cause cancer. People exposed to 1000 mSv have a 3% chance of getting cancer due to the radiation.

e How does the chance of Josy getting cancer from radiation compare with this? Draw a ring round your choice.

more than 3%	about 3%	less than 3%	much less than 3%

f Some people have occupations which increase their radiation dose. Choose three sources of radiation dose that could be increased by a person's job. Write the job in the last column of the table (at the top of the page).

3   How risky is radon?

Radon is a gas which is produced naturally in some rocks. It emits alpha radiation.

a Complete the sentences below. Draw a ring round the correct **bold** words.

Radon gas gives off **alpha** / **infrared** radiation. This is the **most** / **least** ionizing of the radiations. This means that alpha radiation gives a **bigger** / **smaller** dose than the same amount of the other ionizing radiations. Alpha radiation has a **short** / **long** range in air and is **easily** / **never** absorbed. Radon gas presents a **small** / **big** hazard from irradiation because the alpha radiation does not penetrate the skin. However, radon presents a bigger hazard from **contamination** / **evaporation** because it can get inside a person's body.

b The statements below describe why radon is hazardous because of contamination. But they are in the wrong order.

Draw arrows to link the statements in the correct order. The first one has been done for you.

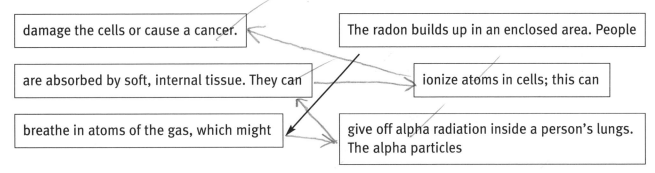

damage the cells or cause a cancer.	The radon builds up in an enclosed area. People
are absorbed by soft, internal tissue. They can	ionize atoms in cells; this can
breathe in atoms of the gas, which might	give off alpha radiation inside a person's lungs. The alpha particles

c You are exposed to radiation in the environment by irradiation and by contamination.

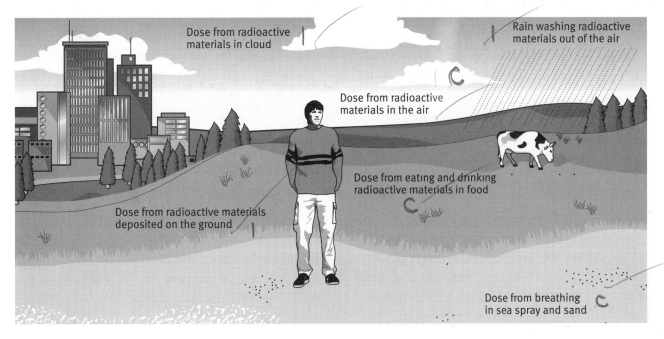

The picture describes some of the sources of radiation you are exposed to. Next to each description, write:

⇨ **I** if you think it is an example of **Irradiation**
⇨ **C** if you think it is an example of **Contamination**

15/17

35

4 Radiation and health – imaging and treatment

a The types of radiation given off by radioactive materials have different properties.
Complete the table by using ✓ and ✗ to record the different properties of the radiation.

Type of radiation	Is absorbed by a thin sheet of paper	Is absorbed by a thin sheet of aluminium	Is absorbed by a thick sheet of lead
alpha	✓	✓	✓
beta	✗	✓	✓
gamma	✗	✗	✓

b Ionizing radiations are used in hospitals. Look at the types of radiation on the left.

i Draw a line from each type to match it with its properties in the middle.

Type of radiation	Properties	Use
X-rays	Penetrates tissue easily and is given out by the nuclei of atoms which can be injected into the body.	Taking photographs of internal structures like bones
Beta radiation	Passes through soft tissue and is absorbed by dense tissue. It can be controlled and directed by the machine that makes it.	Providing images of organs to look for abnormal function
Gamma radiation	Is absorbed by cells and can kill them.	Treating cancers by killing the cancerous cells.

ii Draw a line from each properties box to one of the uses on the right.

c Complete the sentences below. Draw a ring round the correct bold words.

Some **medical** / ~~sporting~~ techniques use ionizing radiations. Like any ionizing radiation, they can damage **cells** / ~~equipment~~ and there is a chance that they will start a **cancer** / ~~virus~~. However, the chance is very **small** / ~~large~~. Most people think that the risk is outweighed by the **benefits** / **costs** of using the technique to investigate or treat a medical problem.

d Jo lives in the UK. She has been feeling unwell and is given an injection of DMSA, which is taken up by her kidneys. It gives out gamma radiation and allows her doctors to get an image of the working parts of her kidneys. The gamma scan shows they are working normally. Joseph lives in Uzbekistan. He has been feeling ill. He is treated with painkillers.

i The gross domestic product (GDP) for the UK and Uzbekistan are €900 and €1800 per person per year. Which figure is the UK's GDP? £900 ✗

ii Why is Joseph not given a gamma scan as a precaution. it costs more too expensive

11/13

5   Radiation and health – protecting patients and staff

a Radiologists regularly work with radioactive sources. It is important to take precautions to minimize the risk from the radiation. Look at the precautions below. Draw a line to match each precaution with the reason for it.

Precaution

| wear gloves and an apron |

| wear protective clothing and stand behind a screen |

| wear a special badge that is sensitive to ionizing radiation |

Reason

| to block out the radiation and reduce the dose and risk due to irradiation |

| to monitor the radiation dose over a year |

| to prevent clothes and skin being contaminated with sources of ionizing radiation |

b The picture on the left shows a type of radiation badge that is worn by radiation workers. The diagram on the right is a cross-section through the badge.

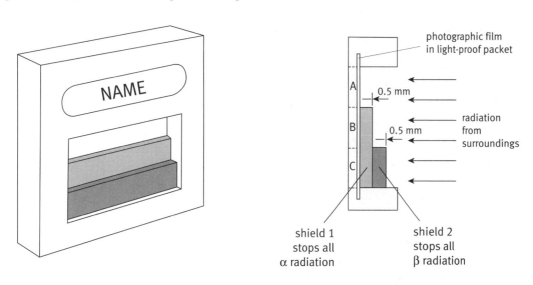

i Put ticks ✓ in the table to show which types of radiation reach each region of the photographic film. Some regions will be exposed to more than one type of radiation.

Radiation Region	alpha	beta	gamma
A	✓	✓	✓ ✗
B	✗	✓	✓
C	✗	✗	✓ ✗

ii Look at these occupations. Put a tick ✓ next to those people who are 'radiation workers'.

- ☑ a hospital radiologist
- ☑ a scientist in a nuclear power station
- ☐ a radio broadcaster
- ☐ a chef in a 'bar and grill'
- ☑ a technician who sterilizes surgical equipment with gamma rays

7/9

48/56

c i There are risks in all walks of life. Here are some people talking about risks.

Draw a line to match each statement with the best explanation.

Statements about risk

Explanations

radiologist

> My maximum-allowed dose is 20 mSv per year, but it is kept well below this level at about 1.5 mSv per year.

> This is because one way of thinking about risk is to include both the chances of it occurring and the consequences if it does occur.

policewoman

> The risk of being mugged is higher than the risk of your car being stolen, even though the chances of them happening are similar.

> This is because the risk depends on the total radiation dose you have received.

doctor

> The risk to your health of not doing the investigation using radiation is greater than the risk from the radiation itself.

> This is an example of the requirement to keep the risk as low as reasonably achievable.

patient

> The doctor wanted to know how many X-ray investigations and radiotherapy treatments I had had in the past.

> Because radiation cannot be seen, some people may think the risk is greater than it is.

recruitment officer for radiologists

> You might be surprised to know that the risk of death per year for construction workers (1 in 16 000) is similar to that for radiation workers (1 in 17 000).

> This is an example of balancing the benefits against the risks.

ii One of the explanations is known as the ALARA principle. Write 'ALARA' under that explanation.

iii Look at the statement by the recruitment officer. For which type of worker do you think

➥ the perceived risk is greater than the actual risk ....

➥ the actual risk is greater than the perceived risk ....

iv Complete this sentence about the policewoman's statement. Draw a ⟨ring⟩ round the correct **bold** words.

The risk of being mugged is **higher / lower** than the risk of having your car stolen because the

reward / consequence is much **worse / better**. Although the chance is the same, risk is a combination of

chance and **consequence / time**.

6   Changes inside the atom – radioactive decay

a Carbon has more than one type of atom. In a sample of carbon, most of the atoms will be carbon-12 atoms, which are stable. However, a tiny proportion will be carbon-11 atoms. These are radioactive.

   i Look at the materials on the left. They all contain carbon. Draw lines to link each material with two or more of the statements on the right. Each statement can be used once, more than once, or not at all. The first line has been drawn for you.

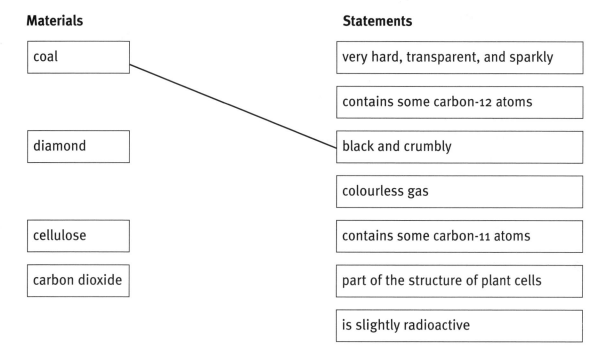

Materials

coal

diamond

cellulose

carbon dioxide

Statements

very hard, transparent, and sparkly

contains some carbon-12 atoms

black and crumbly

colourless gas

contains some carbon-11 atoms

part of the structure of plant cells

is slightly radioactive

   ii Three of the statements apply to all of the materials – they do not depend on the chemical properties of the material. Put a tick ✓ next to these statements.

   iii Complete these sentences. Draw a ring round the correct **bold** words.

   The materials above are all **the same / different**. They have different **chemical / detached** properties.

   However, they all contain some radioactive atoms of the **element / compound** carbon. The radioactive

   carbon **is / is not** affected by the chemical properties of the material. A carbon-11 atom is radioactive

   because its **nucleus / neighbour** is unstable. It is the stability of the **nucleus / lattice** that determines

   the stability of the atom.

b Look at the diagram of an atom on the right. The boxes describe parts of the diagram.

   i Draw a line from each of these boxes to the correct part of the atom. The first one has been done for you. Each part can be linked to more than one box.

   ii If the atom were drawn to the same scale as the nucleus, about what would its diameter be?

   Draw a ring round your choice.

1 cm	10 cm	50 cm	10 m	500 m

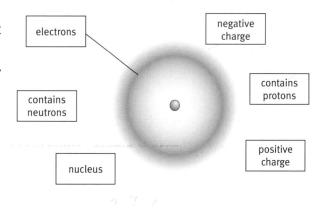

electrons

negative charge

contains protons

contains neutrons

positive charge

nucleus

c The nucleus contains two types of particle: the proton and the neutron. The statements on the left describe either the proton or the neutron. Link each statement with one of the boxes on the right. The first one has been done for you.

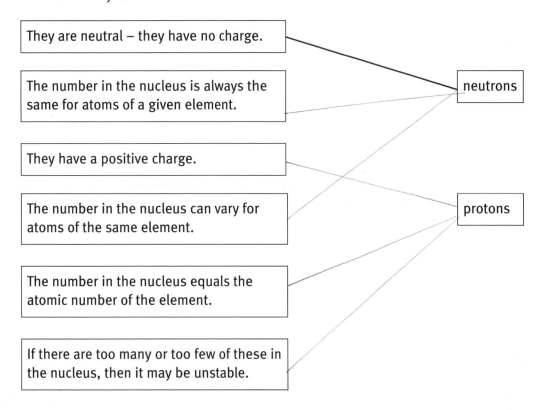

They are neutral – they have no charge.

The number in the nucleus is always the same for atoms of a given element.

They have a positive charge.

The number in the nucleus can vary for atoms of the same element.

The number in the nucleus equals the atomic number of the element.

If there are too many or too few of these in the nucleus, then it may be unstable.

neutrons

protons

d The diagrams below represent four different atoms. The number of protons and neutrons is shown in the nucleus of each one.

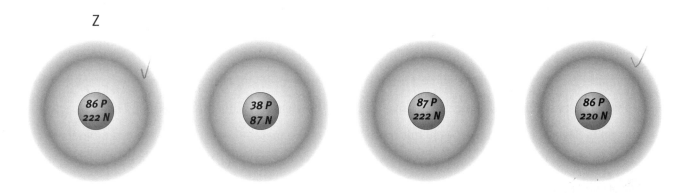

Z

86 P 222 N

38 P 87 N

87 P 222 N

86 P 220 N

i Put a tick ✓ next to the two atoms which are from the same element.

ii Atom Z is unstable and decays by giving out an alpha particle.

It changes into the atom shown on the right.

How can you tell that this is an atom of a different element?

it has 2 less protons.

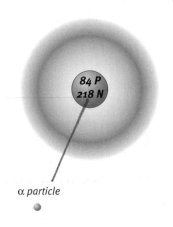

84 P 218 N

α particle

iii Draw an arrow from the atom to the alpha particle to show where the alpha particle came from.

7 Using radioactive materials

a The picture below shows a system that uses a radioactive source to measure the thickness of paper.

Complete these sentences. Draw a ring round the correct **bold** words.

The paper **pulp** / **clips** pass through the rollers and then through a beam of radiation. If the paper becomes too thin, the detector will measure **more** / **less** radiation. If this happens, the **machine** / **supervisor** will adjust the rollers to **reduce** / **increase** the pressure on the paper.

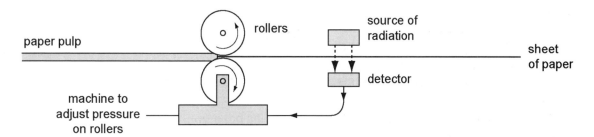

b Food can be irradiated to kill bacteria. This prevents spoilage. Explain why this can be preferable to heating, drying or canning the food.

....

c The picture below shows a smoke alarm. It uses a source of alpha radiation.

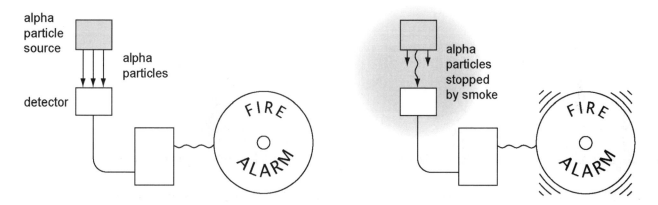

Look at the statements below. They describe what happens if smoke passes into the detector. They are out of sequence. Draw arrows to join the statements in the correct order. The first one has been done for you.

If some smoke gets in the chamber, it	Normally, there is no smoke in the detector. The alpha radiation
changes and that sets off the alarm.	stops some of the alpha radiation so the detector's signal
carry a current in the chamber and the alarm is kept off.	ionizes air in a small chamber. The ions

8 Nuclear power stations

A nuclear power station is similar to a fossil-fuel power station, except that the energy source is a nuclear fission reaction rather than the burning of a fossil fuel.

a Both types of power station consist of a sequence of five parts.

Use these words to fill in the missing parts of this flow diagram.

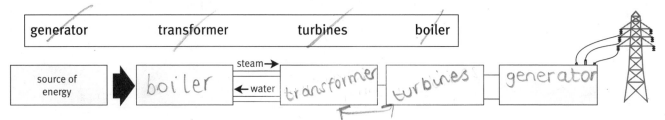

b Some parts of gas and electric power stations are shown on the left. Their roles are described on the right.

Draw a line to match each part with its role. The first one has been done for you.

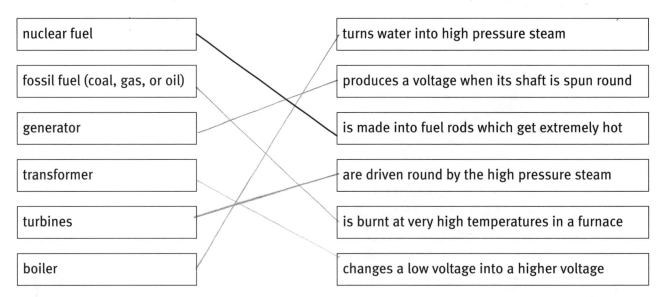

c The table shows some of the effects of using fossil fuels and nuclear fuel in power stations.

Complete the table below by crossing out the statements that do not apply. You should be left with one statement in each box. The first box has been done for you.

	Fossil fuel power stations	**Nuclear power stations**
The power stations produce . . .	~~no wastes~~ carbon dioxide waste ~~radioactive waste~~	~~no wastes~~ ~~carbon dioxide waste~~ radioactive waste
This waste . . .	contributes to global warming ~~is hazardous for thousands of years~~ ~~is completely safe~~	~~contributes to global warming~~ is hazardous for thousands of years ~~is completely safe~~
People living near the power stations are . . .	~~at a risk of catastrophic accidents~~ sometimes exposed to other pollutants ~~unaffected~~	at a risk of catastrophic accidents ~~sometimes exposed to other~~ ~~pollutants~~ ~~unaffected~~

d Complete these sentences. Draw a ring round the correct **bold** words.

Uranium-235 is a **nuclear** / **fossil** fuel. If its nucleus absorbs a **neutron** / **electron**, it becomes extremely **unstable** / **radioactive** and splits into two. This is called nuclear **fission** / **fusion**. When the nucleus splits, it **releases** / **absorbs** energy.

e The diagram shows how uranium-235 can 'chain react'. The statements below describe the chain reaction. They are out of sequence. Draw arrows to join the statements in the correct order. The last statement will point back to one of the earlier statements forming a loop. The first arrow has been drawn for you.

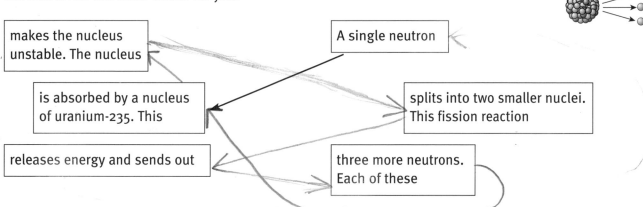

makes the nucleus unstable. The nucleus	A single neutron
is absorbed by a nucleus of uranium-235. This	splits into two smaller nuclei. This fission reaction
releases energy and sends out	three more neutrons. Each of these

f The diagram below shows the nuclear reactor and boiler parts of a nuclear power station. Each time a word is printed in outline, colour it in. Then shade in the corresponding part of the diagram with the same colour.

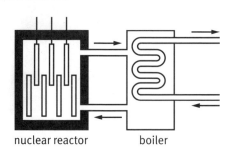

nuclear reactor   boiler

The fission reaction in the fuel rods releases a lot of energy and heats the coolant liquid. The hot coolant then flows around the coiled tube in the boiler and converts water in the tube into steam.

The fission reaction can be slowed down by lowering the control rods.

g Complete the sentences using these two words. You can use each word as many times as you like.

fission   chemical

➜ _fission_ is a nuclear reaction inside an atom, whereas combustion is a _chemical_ reaction between atoms.

➜ In combustion, _chemical_ bonds are made and broken; in _fission_, a nucleus splits into smaller nuclei.

➜ A _fission_ reaction releases about a million times more energy than a _chemical_ reaction because the forces inside the nucleus are so much bigger than the forces between atoms.

9 The pattern of radioactive decay

a Complete these sentences. Draw a ring round the correct **bold** words.

Fission reactions produce a number of **radioactive** / **plastic** products. The activity of these products

decreases / **increases** over time. The time it takes for the activity to drop **by a half** / **to zero** is called the

half-life. Different products **can** / **do not** have very different half-lives. The **longer** / **shorter** the half-life, the

longer a product will be radioactive.

b Two fission products and their half-lives are given in the table:

Fission product	Iodine-131 (A)	Iodine-129 (B)
Half-life	8 days	15.7 million years

Each of these graphs has three points marked on the curve, showing the amount of radioactive decay at different times. For each graph, complete the times on the horizontal axis.

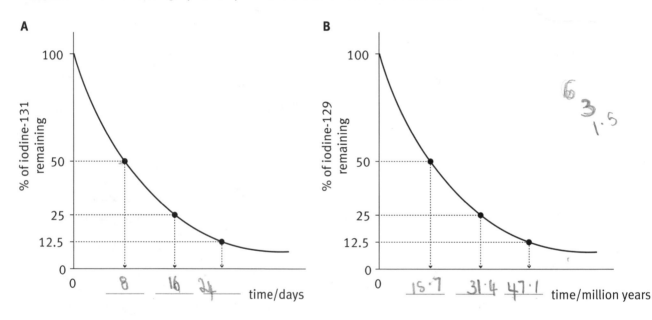

c A sample of radioactive waste contains 64 g of iodine-131.

  i How much iodine-131 will be left behind after:

   ➔ 8 days? __50__   ➔ 16 days? __25__ g   ➔ 24 days? __12·5__ g

 ii How long will it take until the amount of iodine-131 has dropped to 2 g?

   Number of half-lives = ____6____   time = ____95____ days

 iii Two of the phrases below can be linked to describe when the activity of a sample is low enough to be considered safe. Underline the two phrases.

about a tenth of the half-life	about equal to the background count	about ten times the original activity

10 Dealing with radioactive waste

Decisions have to be made about how to deal with existing radioactive waste.
This waste was produced by nuclear power stations and medical uses of radioactive materials.

a Draw a line to match each statement to the type of radioactive waste it best describes.

Made up of the most dangerous fission products from used fuel rods.	 Low-level waste (LLW)	It is very bulky, packed in drums, and kept in specially protected landfill sites.
Made up of used protective clothing, and refuse and rubble with low radioactivity.	 Intermediate-level waste (ILW)	It is not very bulky but it gets so hot that it has to be stored under water.
Made up of materials from inside the reactor. It contains some fission products with very long half-lives, but it does not get hot.	 High-level waste (HLW)	It is chopped up and mixed with concrete, then stored in large steel containers.

b Choose words from this list to complete the sentences below. Each word can be used once, more than once, or not at all.

absorbed	food	hazardous	ALARA	irradiation	safe
transmitted	contamination	precautionary			

Intermediate-level waste is ___transmitted___ for a long time. The radiation it gives off is

___absorbed___ by its packaging so there is very little risk from ___contamination___ *(irridation)*. However, there is a

long-term risk from ___contamination___ *(contamination)*. It must be kept out of water supplies and the ___food___ chain

for tens of thousands of years. Scientists are still looking for a permanent disposal method that they are

sure will be ___safe___ for these lengths of time. No permanent method will be used until they are sure it

is safe; this is an example of the ___precautionary___ *(ALARA)* principle.

11 Electrical power production

The table below shows the main costs associated with building, running, and decommissioning a nuclear power station.

a The costs are assessed for its whole lifetime – from the cradle to the grave.

What is this type of assessment called:

L _Ower_ C _ase_ A _ssesment_

b Add together rows F and G to find the total running cost for the lifetime of the power station.

Total running cost = _4 40_ + _700_

= £ _1200_ _(140)_ million

Put your answer in row H.

	Type of power station	Amount
A	Building cost / £ millions (£m)	3000
B	Typical lifetime / years	40
C	Power output / MW	1000
D	Lifetime energy output / million kWh	~~4400~~ 350 400 354800
E	Fuel cost per kWh produced / p	1.2 354801.2
F	Total fuel cost / £m	440 354 01.2
G	Lifetime maintenance cost / £m	700
H	Lifetime running cost / £m	
I	Decommissioning cost / £m	3900
J	Total (cradle to grave) cost / £m	
K	Cost per kWh produced / p	3.4

The flow chart shows the life of a nuclear reactor.

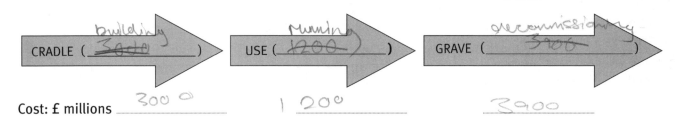

CRADLE (_building_ ~~3000~~)   USE (_running_ ~~1200~~)   GRAVE (_decommissioning_ ~~3900~~)

Cost: £ millions _300 0_   _1 200_   _3900_

c Fill in the gap in each arrow with one of these labels: **running**   **decommissioning**   **building**

d Put the cost of each stage under each arrow.

e Which cost is the highest for a nuclear power station? <u>Underline</u> the correct answer.

building	running	decommissioning

f Add up rows A, H, and I to find the total cost for the power station (from cradle to grave).

Total lifetime cost = _3000_ + _1240_ + _3900_ = £ _7040_ million
Put your answer in row J.

g Row K shows the total cost per kWh of energy produced. Explain why this is more than the fuel cost per kWh in row E. _The fuel cost is cheaper to produce but cost is easer to produce £._

12 Energy futures

a The table shows the Life Cycle Assessment costs for three methods of generating electricity.

i Which type of power station has the cheapest source of energy?

Gas

ii Gas costs more than nuclear fuel. Nevertheless it is cheaper to use gas to generate electricity. Explain why.

More energy available per kWh

iii The lifetime energy output of a wind farm is lower than expected (*). This is because a wind farm rarely runs at its maximum power output. Explain why.

Amount \ Energy source	Wind farm	Gas	Nuclear
Building cost / £ millions (£m)	50	410	3000
Typical lifetime / years	20	40	40
Power output / MW	60	600	1000
Lifetime energy output / million kWh	4000*	210 000	350 400
Fuel cost per kWh / p	0	1.6	1.2
Lifetime maintenance cost / £m	38	450	700
Decommissioning cost / £m	40	480	3900
Cost per kWh produced / p	3.2	2.3	3.4

b Some other issues are shown in the table below. Put ticks ✓ in the boxes to show which generation method they apply to. You can put one or more ticks in each row. The first row has been done for you.

Issue \ Generation method	Wind farm	Gas	Nuclear
It takes ten years to build and commission a new power station.			✓
It requires a large area of open land.	✓		
It is built of concrete, which releases carbon dioxide.		✓	✓
The energy source is free and sustainable.	✓		
Produces significant amounts of carbon dioxide for each unit of electricity, as it is generated.		✓	
The supply cannot be guaranteed – it depends on the weather.	✓		
The energy source is likely to become scarce and expensive.		✓	
It produces a guaranteed constant supply without releasing any carbon dioxide.			✓
Decommissioning is a long and expensive process. The waste and the buildings are all radioactive.			✓
The structures can be dismantled with relative ease.	✓		

c Use part **a** and **b** make a proposal about how to proceed with one of the energy sources.

Wind farms are costly to set up but have the benefits of less damage to the environment.

1 Interactions and forces

Forces happen because of an **interaction** between two objects. They happen in **pairs**.

a In each of the bullet points below, draw a ring round the correct **bold** word.

The two forces in an interaction pair

➜ are **always** / **sometimes** / **never** the same size

➜ act in **the same** / **random** / **opposite** directions

➜ act on **the same** / **a different** object

b In each of these drawings, add an arrow in red to show the named force.

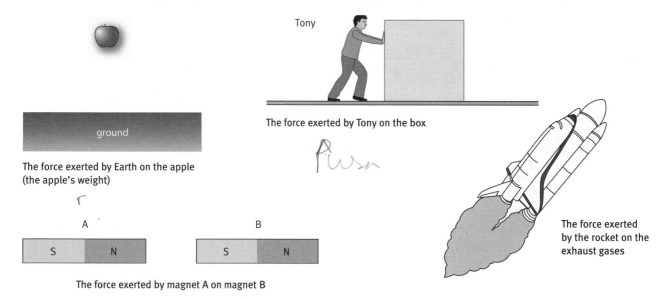

Tony

The force exerted by Tony on the box

Push

ground

The force exerted by Earth on the apple
(the apple's weight)

A   B

| S | N | | S | N |

The force exerted by magnet A on magnet B

The force exerted
by the rocket on the
exhaust gases

c Complete these sentences by filling in the missing words. Each sentence will help you with the next one.

i Magnet A pulls magnet B to the left; magnet B pulls magnet B to the _____right_____ with the same size force.

ii Tony pushes the box to the right; the box exerts a force on Tony. It pushes him to the

_____left_____ with the _____same_____ size force.

iii The Earth exerts a downwards force on an apple; the apple exerts an _____upwards_____ force on the

_____earth_____. It pulls back with the _____different size_____ _____force_____.

iv The Shuttle exerts a force backwards on the exhaust gases; the exhaust gases

_____

d Now add a second arrow to each of the pictures above. This should show the second force in the interaction pair. Use a different colour and label the force like this:

'force exerted by _____kid_____ on _____door_____,

2   Getting moving

a Look at the picture on the right.
It shows Sophie pulling on a rope
which is attached to a wall.

Sophie

Draw a circle round two places where
friction is helping her grip.

b A bicycle tyre grips the road. When the wheel turns, the bottom of the tyre pushes backwards on the
ground. In the picture, the tyre exerts a force to the left on the ground.

turning wheel

i   What is the force that allows the tyre to
push backwards on the ground? _friction_

ii The ground exerts a force on the tyre.
What is the direction of this force? _Pull_

iii Draw an arrow on the wheel to show the direction
and size of this force.

iv Label the two arrows on the picture.

c Use words from the list to complete the sentences below.

Words may be used once, more than once, or not at all.

upwards	downwards	forwards	backwards	grip	rope	exerts

➜ A racing car accelerates at the start of a race: friction allows the tyres to ___grip___ the track.

The tyres exert a force backwards on the track; the track ___exert___ a force ___backwards___

on the tyres. This makes the car speed up.

➜ A climber uses a rope to go up a mountain: friction allows her hands to grip the ___rope___.

Her hands exert a ___upwards___ force on the rope; the rope ___downwards___ an

___upwards___ force on her hands.

➜ Mick is trying to walk to the shops. The ground is icy and slippery.

His feet cannot get a ___grip___ to push ___forwards___ on the surface.

So the surface does not push him ___backwards___.

3 How surfaces hold things up

a The picture shows a bag on a springy cushion.

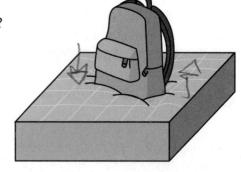

i How can you tell that the bag is exerting a force on the cushion?

It is lying on a solid surface

ii Draw a red arrow to show the force that the bag exerts on the cushion. Label this arrow 'force exerted by bag on cushion'.

iii The cushion reacts by pushing up on the bag.
How does the cushion exert a force on the bag?

Upthrust Pushes the bag up.

iv What is the name of this force? Upthrust.

v Draw a second arrow to show the force the cushion exerts on the bag. Use a different colour and label this force with its name.

vi What would happen to the cushion if the bag were removed?

It would retain its normal size

b The picture shows an apple resting on a table. Its weight is pulling it down. But the apple is not falling.

i What is the force (and object) that stops the apple from falling?

Gravity from the table.

ii Does the table get squashed by the apple? .

iii Explain why we don't see the table being squashed.

The table has a larger surface area and weighs more

iv Explain how the table exerts a force on the apple.

The apple Because of its sturdy surface

c Complete this sentence to describe the forces between an object and a surface it is resting on.

When an object rests on a surface, it exerts a force Gravity on the surface.

The surface exerts a force friction on the object.

This is called a Upthrust force.

4 Weight and reaction

So far, you have looked at pairs of forces acting *between* two different objects.
Now you will work with more than one force acting on a *single* object.

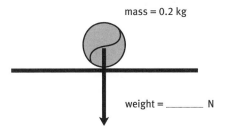
mass = 0.2 kg

weight = _____ N

a The picture shows a tennis ball on the ground. Its mass is 0.2 kg.

  i What is its weight? (Take the gravitational field strength as 10 N/kg.)

  weight = _____ N

  ii Write this value on the picture.

  iii Draw an arrow to show the reaction force exerted by the ground on the ball.

  iv Label this force.

  v There are two forces acting on the ball.

  What is the **resultant** force acting on the ball? _____

b The picture shows a mass hanging on a thread. Its weight is 4 N.

  i Draw arrows for the two forces that are acting on the mass.

  Use these phrases to label the forces:

  ➔ force exerted by the Earth on the mass
  ➔ force exerted by the thread on the mass

  ii What is the resultant force on the mass? _____

  iii Imagine the thread breaks. Put a cross ✘ next to the force that will disappear.

  iv Put a tick ✔ next to the force that will still be there after the thread breaks.

  v Immediately after the thread breaks, what is the resultant force on the mass?

  _____

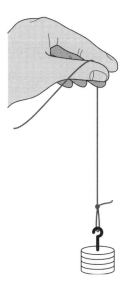

c Clare and Sophie have a tug-of-war. They pull the rope with the same amount of force.
They are not moving.

  i Draw and label arrows on the diagram to show:

  ➔ force exerted by Clare on the rope
  ➔ force exerted by Sophie on the rope

  ii What is the resultant horizontal force acting on the rope?

  _____

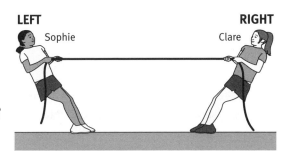
LEFT   RIGHT
Sophie   Clare

5 Average and instantaneous speed

a The average speed of an object is:

$$\text{speed (m/s)} = \frac{\text{distance travelled (m)}}{\text{time taken (s)}}$$

In the 1995 World Athletic Championships, Linford Christie was the defending champion, in the 100 m sprint. He lost to Donovan Bailey, who won with a time of 9.97 seconds.

Calculate Bailey's average speed for the race.

Bailey's average speed = ... m/s

b The table below shows the time six of the athletes took to complete the race.

Finishing position	Athlete	Finishing time in s	Average speed in m/s
1	Bailey	9.97	
2	Surin	10.03	9.97
3	Boldon	10.03	
4	Fredericks	10.07	
5	Marsh	10.10	
6	Christie	10.12	

   i Complete the table by calculating the average race speed for the athletes. The second one has been done for you.

  ii What happens to the speed as the time gets longer? ....

c i How long would it take for Surin to run 800 m at this average speed?

     time for Surin to run 800 m = ... s

  ii Complete these sentences by drawing a ring around the correct **bold** words.

In reality, if Surin were to keep running for 800 m, his average speed would be **more / less** than 9.97 m/s. So the time he would take is **longer / shorter** than the answer above.

Where speed is changing, the average speed and instantaneous speed can be different.

d The following diagram shows how the positions of six of the runners changed at each 25 metre mark.

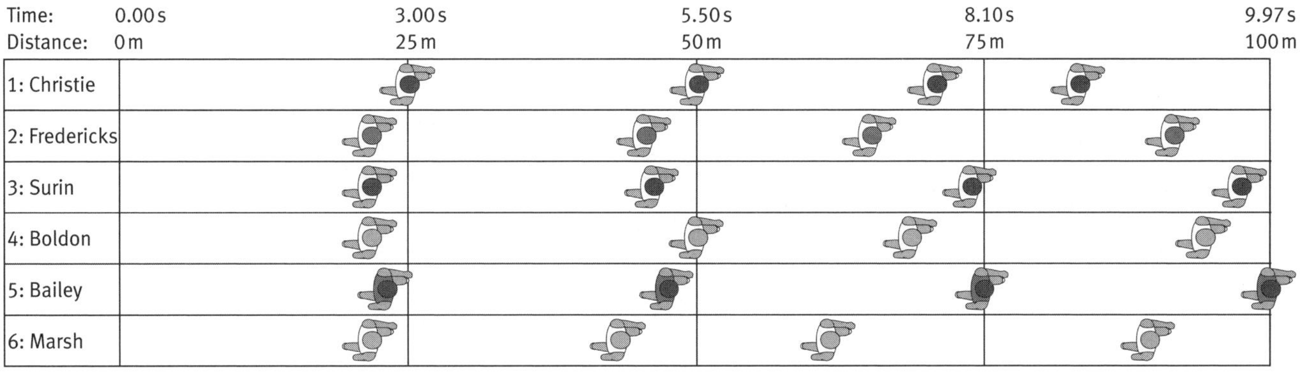

Time:	0.00 s	3.00 s	5.50 s	8.10 s	9.97 s
Distance:	0 m	25 m	50 m	75 m	100 m
1: Christie					
2: Fredericks					
3: Surin					
4: Boldon					
5: Bailey					
6: Marsh					

i Which athlete got the best start to the race (at 3.00 s)? _____

ii Use the information in the diagram to calculate Bailey's average speed over the last 25 m of the race.

Bailey's average speed for last 25 m = _____ m/s

iii Complete these sentences using words from the box. There is one word you don't need to use and one word you need to use twice.

instantaneous	average	lower	higher	speed

Bailey's speed at any one moment is called his _____ speed. Just after the start,

his instantaneous _____ is lower than it is when he gets going. As a result, his

_____ speed for the last 25 m is _____ than his _____

speed for the first whole race.

e During the race, Christie suffered a hamstring injury.

Look carefully at the pictures and decide where you think Christie's injury happened. Explain your answer.

_____

_____

6 Using graphs to summarize motion

a Look at the distance–time graph on the right.

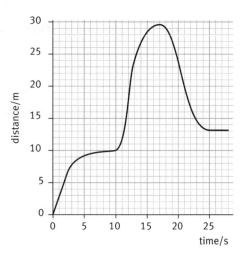

i Label the graph with these letters to show where the object is

   A moving away from the start with a constant speed

   B moving backwards towards the start with a constant speed

   C stationary

ii You can tell from the graph that the object moved faster between 0 and 5 seconds than it did between 10 and 15 seconds. Explain how you can see this from the graph.

....

iii Work out the speed of the object between 10 seconds and 15 seconds.

speed = ... m/s

Straight slopes on a distance–time graph tell you that an object is moving at a constant speed.

Curves tell you that an object's speed is **changing** – it is speeding up or slowing down.

b Label the graph with these letters to show where the object is

   D slowing down (going forwards)

   E speeding up (going forwards)

   F stationary

   G slowing down (going backwards)

   H speeding up (going backwards)

A graph of speed against time is another useful way of showing how an object moves.

c Look at the distance–time graphs (left) and the speed–time graphs (right) below.

Draw lines to match each distance–time graph with its corresponding speed–time graph. The first one has been done for you.

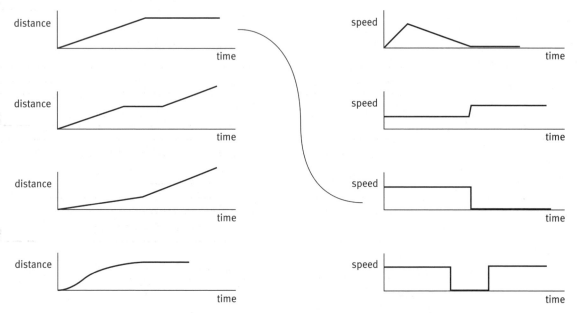

A tachograph is a speed–time graph to record a lorry's journey.

There are strict rules to control how long a lorry driver can drive without taking a break. A tachograph is used to record the lorry's journey.

d i What was the highest instantaneous speed of the lorry?

....

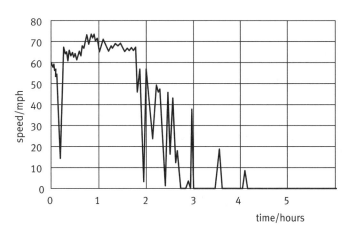

ii During the journey the lorry

A drove in a town

B drove on the motorway

C parked up for the night

D made some deliveries

Use the letters to label sections of the graph that show these actions.

iii The average speed of the lorry in the first two hours was 62 mph.

Draw a coloured line to show this on the graph.

iv In the first two hours, mark a time where the instantaneous speed was much less than the average speed.

v Suggest what might have happened at that moment. ....

A tachograph shows the **speed** of a lorry. It is a speed–time graph. You cannot tell which direction it was going.

A **velocity–time** graph gives a fuller description. It shows the **speed** *and* **direction**. Velocity means the speed in a certain direction.

e The velocity–time graphs below show the motion of a car travelling along a straight road running north–south.

Draw lines to match each graph with the best description of the motion. One has been done for you.

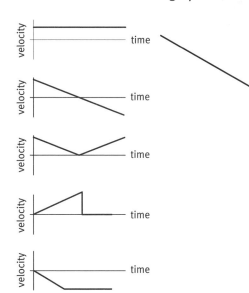

The car travels north, steadily slowing down until it stops momentarily and then gradually speeding up again.

The car is travelling north at a constant speed.

The car gradually speeds up as it travels north and then suddenly comes to a stop.

The car gradually speeds up as it travels south, eventually travelling at a constant speed.

The car, which is travelling north, slows down and comes to a stop. It then travels south, getting faster and faster.

f Complete this sentence: When the car is moving backwards, its velocity is ....

7 Momentum

momentum = mass × velocity
(kg m/s)   (kg)   (m/s)

a Look at the three balls on the right.
They are travelling at different speeds.

i Calculate the momentum of the golf ball.

ii Use the same method to calculate the momentum of the tennis ball and football.

iii Put a tick ✔ next to the ball with the biggest momentum.

iv Each of the balls strikes a large skittle. Put a cross ✘ next to the ball which is most likely to knock it over.

v Explain your choice in part **iii**.

....

....

....

....

0.05 kg

→ 30 m/s

momentum = mass × velocity

= .... × ....

= .... kg m/s

0.05 kg

→ 45 m/s

momentum = mass × velocity

= .... × ....

= ....

0.5 kg

→ 4 m/s

momentum = ...

b Jo is ice skating. To get her moving, Sam gives her a push.

Work out Jo's momentum after she is pushed.

Momentum = ... × ....

= .... kg m/s

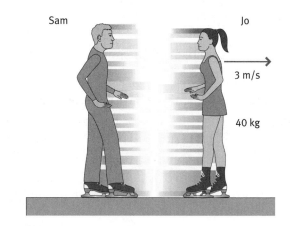

Sam   Jo

→ 3 m/s

40 kg

c Use words from the list to complete the sentences below.
Words may be used once, more than once or not at all.

zero	momentum	force	left	right	same	opposite

Jo is stationary before Sam pushes her. Her momentum is .... Sam exerts a

.... on Jo, pushing her to the .... The force acts for a short time;

as a result, her .... changes. She gains momentum going to the right – the

.... direction as the force that made this happen.

8 Forces and change of momentum

a Sam and Jo are the skaters from question **7b**.

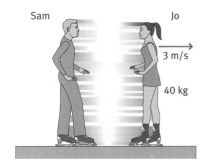

Before Sam pushes Jo, she is standing still. Her speed is 0 m/s.
Sam exerts a force on Jo which increases her momentum.
Sam could increase Jo's momentum more if he

⇢ pushes her with more **force** *or*

⇢ pushes her for a longer **time**

Write in the missing quantities and units to complete
the equation:

Change of momentum = ... × ....

(......................................)   (......................) (......................)

b Jo gains 120 kg m/s of momentum when Sam pushes her. Sam exerts a force on her while they are in
contact. This lasts for 2 seconds.
What is the size of the average force that Sam exerts on Jo?

Sam's force = ... N

c Lucie and Richard are on their skateboards.
The ground is horizontal and they are not moving.

Lucie has a heavy ball. Lucie throws the ball and
Richard catches it.

i This sentence describes what happens to Richard
and Lucie. Put a ⟨ring⟩ around the correct **bold** words
to complete the sentences.

Richard moves to the **left** / **right**; Lucie moves to the **left** / **right**.

ii Look at these phrases. They describe the exchange in terms of forces. But they are out of sequence.
Put numbers in the boxes next to each phrase, to make sentences that show the correct sequence.

☐ To catch the ball, Richard exerts a force on it.

☐ the ball exerts a force on Lucie to the left.

☐ the force pushes the ball to the left.

☐ the force pushes the ball to the right;

☐ The ball exerts a force on Richard to the right.

☐ To throw the ball, Lucie exerts a force on it:

9 Changing momentum safely

a Look at the pictures of two crashed cars. Both cars were travelling at 20 m/s when they hit the wall. Look at these four phrases. They refer to the force on the driver as he is brought to rest.

i Draw lines to link each of the phrases with one of the pictures.

small time

big time

small force

big force

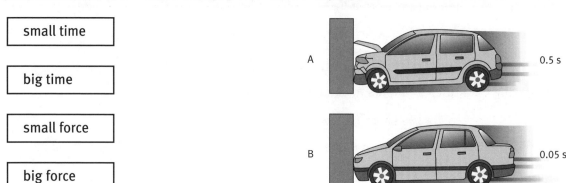

ii The driver's mass is 60 kg.

What is his momentum before the collision?

driver's momentum = ... kg m/s

iii He is brought to rest by the force from the seatbelt. The times are shown in the pictures.

Use the equation on page 42 to calculate the force on the driver in

→ car **A**

→ car **B**

iv It took ten times longer to stop the driver in car A. What effect did this have on the force acting on him?

....

b Imagine the driver in car B had not been wearing a seatbelt.

i What object (or objects) would have provided the force to slow him down?

....

ii It is safer to have *both* a seatbelt *and* a crumple zone. Explain the disadvantages of

→ a seatbelt on its own: ....

→ a crumple zone on its own: ....

10 Steady motion requires no (resultant) force

This is the **first law of motion**:

> If the resultant force acting on an object is zero, its momentum will stay the same.

a Complete these sentences. Draw a (ring) round the correct **bold** words.

This means that a **stationary / weightless** object will stay where it is.

It **will / will not** start moving if there is no resultant force.

A moving object will carry on at **the same / a lower** speed.

In both cases the momentum **does / does not** change because the resultant force is **zero / small**.

b The picture shows Georgia on her bicycle. Imagine Georgia

increases her driving force to 150 N.

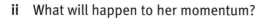

i What will be the resultant force be on the bike?

_____ N to the _____.

ii What will happen to her momentum?

It will _____ to the _____.

c Complete this sentence, which is part of the **second law of motion**:

> When a _____ force acts on an object, its _____ will change in
>
> the _____ direction as the force.

d The pictures show forces on some toy cars. Cars A and B are stationary. The others are moving to the right.

Fill in the blanks for the resultant force and its direction. Then draw lines to match each of
pictures with a correct description of the car's motion. One has been done for you.

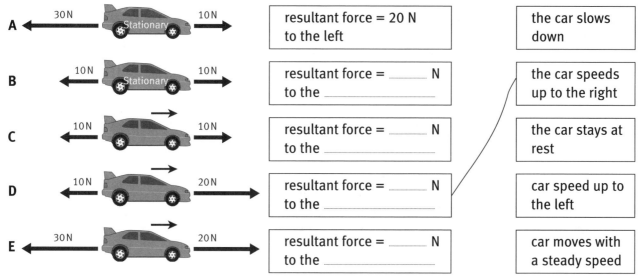

11 Work and change of energy

Whenever you move an object, you have to do some work.

The amount of work depends on

⮞ the force you use

⮞ the distance the object moves

a The equation for the amount of work done is:

work done = force × distance

(_____) (_____) (_____)

Des

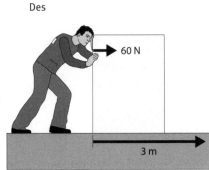
60 N

3 m

i Complete the equation by putting the units in the brackets.

ii Des pushes a box 3 m across the floor. How much work does he do?

work done = _____ N × _____ m

= _____ J

iii Energy is always conserved.

Complete these sentences to describe what happens to the work that Des does.

When Des pushes the box across the floor, the bottom of the box _____ up. The work goes

into _____ the temperature of the box and the ground. When work is done on an object

(or collection of objects), it increases the amount of _____ stored in them.

b Gravitational potential energy (gpe)

Whenever you lift an object upwards, you have to do some work.
You are lifting against the force of gravity, which is pulling downwards on the object.

The object then has the potential to do the work back for you.

i Complete this sentence.

As a result of lifting the object up, it gains gravitational _____ energy (gpe).

ii You lift a box that weighs 200 N on to a table that is 0.5 m high.

How much gravitational potential energy (gpe) has the box gained?

gpe gained = _____ N × _____ m

= _____ J

iii Write down the equation you use to work out change in gravitational potential energy. Include the units.

Change in gpe = _____ × _____

(_____) (_____) (_____)

12 Kinetic energy

a Complete this sentence.

You can do work on an object to make it speed up.

As a result, the object gains _____

_____ (KE).

Chris

mass of Chris
and bicycle = 110 kg

speed = 8 m/s

distance travelled while
speeding up = 50 m

forward force exerted
by Chris = 100 N

b Chris does work at the start of a race to get her
bike going.
From a standing start, she pedals along a straight
road and reaches a speed of 8 m/s.

The mass of Chris with her bike is 110 kg.

i Write down the equation you can use to work out **kinetic energy**. Include the units.

kinetic energy = ½ × _____ × _____

(_____)   (_____) (_____)

ii Calculate the kinetic energy of Chris and her bike at 8 m/s.

kinetic energy =

kinetic energy of Chris and her bike = _____ J

iii Chris produced a driving force of 100 N. Calculate the amount of work she did.

work done = _____ × _____

work done by Chris = _____ J

iv Complete these sentences:

Chris did _____ joules of work; she gained _____ joules of _____ energy.

This is _____ than the amount of work she did to get the bicycle moving.

Remember this important principle: **Energy is always conserved**.

v In which case, explain what happened to the extra work that Chris did.

_____

_____

_____

c i Complete these sentences to describe the energy change of a falling object.

When an object falls it loses _____ _____ energy but

it gains _____ energy.

You can use this idea to work out the speed of the rollercoaster car at the bottom of the hill.

ii Write down the equation to work out how much gravitational potential energy the car loses.

_____

iii Calculate how much gravitational potential energy (gpe) the car loses.

gpe lost = _____ J

Energy is always conserved. Assume that all the gpe lost is gained as kinetic energy.

iv Write down how much kinetic energy the car has gained.

kinetic energy gained = _____ J

v kinetic energy = ½ mass × (velocity)²

Calculate the size of the car's velocity.

car's speed = _____ m/s

This is the speed you might expect it to have.

vi In reality, how will the speed of the car compare with your answer to part v? _____

vii Explain why.

_____

1 Static

Insulating materials get charged up when they are rubbed. This is due to stationary charge. It is called static electricity.

a A piece of polythene is rubbed with a cloth and gets charged. Another piece of polythene is rubbed with the cloth.

   i Complete these sentences. Draw a ring round the correct **bold** words.

     The two pieces of polythene will get **the same / different** charge. This means that they will **attract / repel** each other. The polythene was charged because of the movement of **electrons / ions** which have a **positive / negative** charge.

   ii Look at the picture below. It shows that the polythene became negatively charged. There is a line that shows the transfer of electrons. Draw an arrow on one end of this line to show the direction in which the electrons moved.

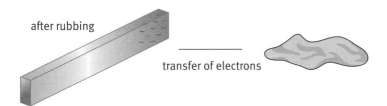

after rubbing

transfer of electrons

   iii Explain how the neutral cloth has become positively charged. ....

....

   iv Draw a line to match the start of each sentence with its correct ending. These sentences summarize the forces between different charges.

Like charges . . .		. . . attract each other.
Opposite charges . . .		. . . repel each other.

b Polythene gets a negative charge when it is rubbed. Perspex gets a positive charge.

   i Complete these sentences. Draw a ring round the correct **bold** words.

     A piece of perspex will **attract / repel** another piece of perspex. A piece of perspex will **attract / repel** a piece of polythene.

   ii Polythene repels nylon. What charge must nylon have? ....

   iii What effect will Perspex have on nylon? ....

c The dome of a Van de Graaff generator gets charged up when it is switched on. Yasmin is holding the dome and her hair stands on end.

   i Explain why her hair stands on end.

....

   ii The teacher discharges the dome and Yasmin's hair falls back down. Rearrange the word below to describe Yasmin's charge once her hair has fallen back down.

                                  unrealt ....

2   Current in a series circuit

a Physicists think that an electric current is the flow of charge – the same charge that causes static electricity. The statements below explain one piece of evidence for this belief. But they are out of sequence. Use arrows to join up the statements in the correct order. The first one has been done for you.

> This implies that the electric current,

> When the spark jumps across the gap,

> These are the same charges that jump across the gap to make a spark.

> is a movement of charges.

> which makes the lamp light up,

> the lamp lights up.

b The picture shows a model of an electric circuit. Bethan is the 'battery'. She pushes the loop of rope around in the direction shown by the arrows. The other pupils let it pass through their hands.

Join the boxes below to show how the model helps to explain an electrical circuit. The first linking line has been done for you.

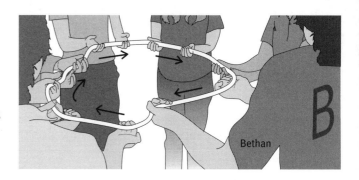

Bethan

Rope circuit

> When Bethan first pushes the rope, it starts moving through everyone's hands at the same time.

> Bethan gets tired after pushing the rope around.

> The others feel their hands getting hot.

> If any one of the others grips the rope firmly, the rope stops moving.

> At any time, the amount of rope leaving each child's hand is the same as the amount going in.

Electrical circuit

> Stored energy is transferred out of the battery.

> Putting in an insulator stops the flow of charge.

> The current is not used up. It is the same everywhere.

> Charge moves throughout the circuit as soon as it is connected up.

> The battery does work on all other components in the circuit.

c All matter is made of atoms, which contain protons (positive) and electrons (negative). In metals, the electrons are free to move throughout the metal. That is why metals are electrical conductors.

The flow of charge around a circuit is called electric current. Traditionally, physicists and engineers use 'conventional current', showing the direction that positive charges would be flowing. The electrons actually flow in the other direction.

i Look at the descriptions below. Draw lines to match current with its description in the middle and its direction on the right.

| 'conventional current' | flow of negative electrons | from – terminal to + terminal |

| actual movement | imaginary flow of positive charges | from + terminal to – terminal |

ii The diagram shows atoms in a wire.

➲ Label the end of the wire joined to the positive (+) battery connection.

➲ Draw an arrow next to the wire to show the direction of 'conventional current'.

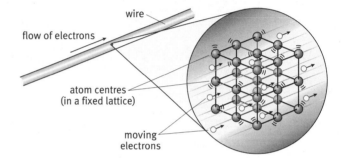

d i This diagram shows a computer model of an electric circuit.

Complete the sentences.

The electric current is being measured by

an _____.

The meter reading is 10 milliamps (mA). This is the same as

_____ amps (A).

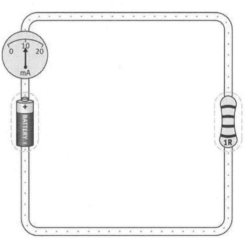

ii This diagram shows the same circuit, but it now has two extra meters.

Put arrows in each meter to show the readings, then complete the sentences.

The charges in an electric circuit never get used up.

Current is the *flow* of charges, so the current is

_____ all the way round the circuit.

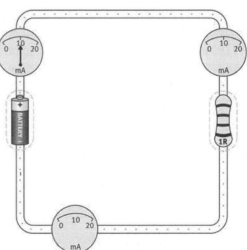

3   Current in parallel circuits

When several components are connected in parallel directly to a battery

    X ➔ the current through each component is the same as if it were the only component present

    Y ➔ the total current from (and back to) the battery is the sum of the currents through each of the parallel components

Circuit 1 on the right shows a single resistor connected to a battery.

The current through the resistor is 6 mA.

a What is the current coming from the battery?

....

b Put arrows in the other two ammeters to show their readings.

In circuit 2, another similar resistor has been added in parallel.

c What current will the current be in resistor A in circuit 2?

....

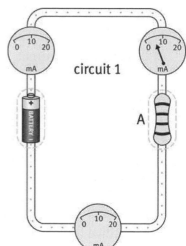

d Which of the statements at the top of the page tells you this?
Draw a ⬭ring⬭ round the correct statement below.

    statement X   **statement Y**

e Draw arrows in ammeters 1 and 2 to show their readings.

f Resistor B is exactly the same as resistor A. What will the current be in resistor B?

....

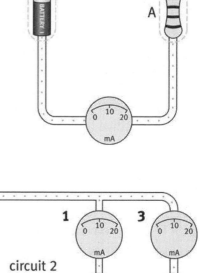

g Draw arrows in ammeters 3 and 4 to show their readings.

h What will be the current through the battery?

.... mA

i Which of the statements at the top of the page tells you this?
Draw a ⬭ring⬭ round the correct statement below.

    statement X   **statement Y**

j Draw an arrow in each of the ammeters 5 and 6 to show their readings.

k Now imagine you add a third resistor in parallel. Put a tick (✓) next to the true statement below.

    ➔ The current through the third resistor will be the same as the currents through resistors A and B and the current coming from the battery will increase.

    ➔ The current coming from the battery will stay the same and the current in resistors A and B will drop slightly.

4 Resistance

a Look at the statements on the left below. Some of them apply to electric current and some apply to electrical resistance.

Draw a line to link each statement with the correct box on the right. The first one has been done for you.

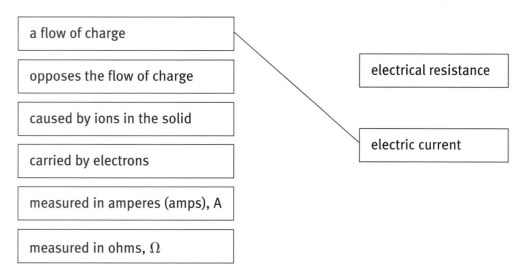

| a flow of charge |
| opposes the flow of charge |
| caused by ions in the solid |
| carried by electrons |
| measured in amperes (amps), A |
| measured in ohms, Ω |

electrical resistance

electric current

b Complete the sentences below. Draw a ring round the correct **bold** words.

➔ The resistance of a circuit determines the **size / direction** of the current.

➔ A big resistance makes it **more / less** difficult for the charge to flow and leads to a **big / small** current.

➔ It is easier for charge to flow through a **smaller / bigger** resistor so the current through it is **smaller / bigger**.

➔ Insulators have a very **small / big** resistance and so the current through them is practically zero.

c Combining resistors in series and parallel will make a new resistance.
   i Complete the statements below (left). Draw a ring round the correct **bold** words.

   ii Draw a line to link each statement with the correct explanation on the right.

| Two resistors in series have a **bigger / smaller** resistance than either one on its own. |

| Because there are more paths that the moving charges can take. |

| Two resistors in parallel have a **bigger / smaller** resistance than either one on its own. |

| Because the moving charges have to pass through one then the other. |

d Look at the networks of resistors below. Each resistor has the same value. Put them in order of increasing resistance. Number them from 1 to 5: the smallest has the number 1 and the largest is 5. The first one has been done for you.

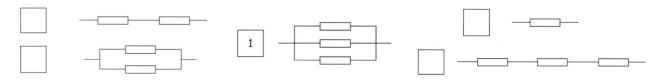

5 Ohm's law

a Use these words to complete the sentences below. Words can be used once, more than once, or not at all.

voltage	size	bigger	smaller	double	battery	proportional

A _____ pushes charge around an electric circuit. A battery's _____ is a measure

of its push. The bigger a battery's voltage, the _____ its push on the charge. In turn, this will

lead to a _____ current. The current is _____ to the voltage. This means

that doubling the voltage will _____ the current.

b Look at the circuit on the right. It has two 1.5 V batteries giving a total voltage of 3 V. The current in the circuit is 24 mA.

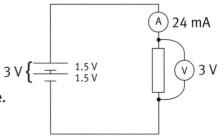

Now look at the circuits below. The resistance of each circuit is the same. However, the number of 1.5 V batteries changes.

i In each circuit, write the voltage across the resistor next to the voltmeter

ii In each circuit, write the current next to the ammeter.

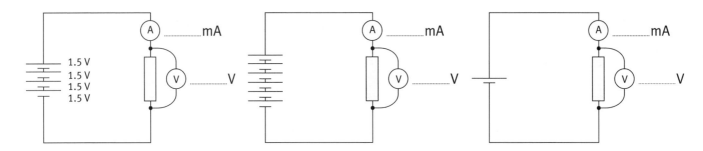

c The graph on the right shows how the current varies in a fixed resistor as the voltage is increased.

i Complete this sentence by filling in the missing word.

The current is _____ to the voltage.

ii Imagine the test is done on a resistor with twice the resistance.

What difference will this make to the currents at each voltage?

_____

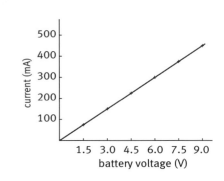

iii Draw in the line you would get for current against voltage with twice the resistance.

d The equation below is used to calculate resistance. Complete the equation by putting in the units.

$$\text{Resistance } (\underline{\hspace{2cm}}, \Omega) = \frac{\text{voltage } (\underline{\hspace{2cm}}, V)}{\text{current } (\underline{\hspace{2cm}}, A)}$$

6   Resistance of LDR and thermistor

a i Some electrical components have a resistance that changes. Look at the three components below left.

➜ Draw a line from each component to match it to the description of how to change its resistance.

➜ Draw a line to join each of the descriptions to a possible use on the right.

The first one has been done for you.

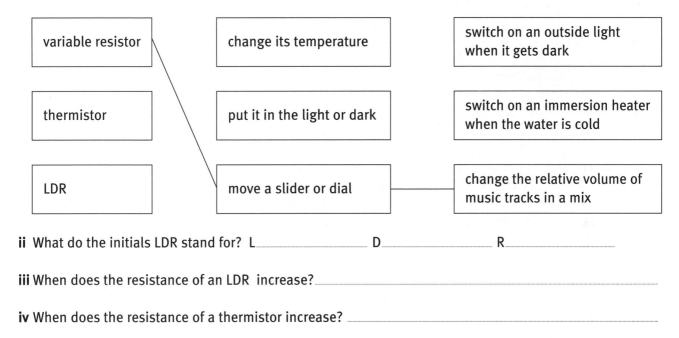

variable resistor	change its temperature	switch on an outside light when it gets dark
thermistor	put it in the light or dark	switch on an immersion heater when the water is cold
LDR	move a slider or dial	change the relative volume of music tracks in a mix

ii What do the initials LDR stand for? L............................ D............................ R............................

iii When does the resistance of an LDR increase?...

iv When does the resistance of a thermistor increase? ....

b Use the names in the list below to label the circuit symbols.

single battery   power supply   filament lamp   switch   LDR   resistor
variable resistor   thermistor   ammeter   voltmeter

7 Potential difference

The voltage of a battery is the 'push' the battery gives the charge. But it is also the work the battery does in pushing each unit of charge around the circuit. Inside the battery, the chemical reactions give each charge some **potential energy**.

The water model can help to show this. The pump pushes on the water, raising it up and increasing its potential energy. It loses this potential energy as it flows round the circuit.

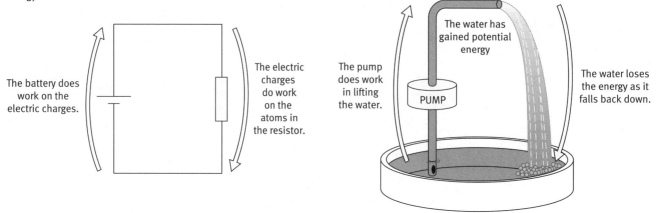

The battery does work on the electric charges.

The electric charges do work on the atoms in the resistor.

The pump does work in lifting the water.

The water has gained potential energy

The water loses the energy as it falls back down.

PUMP

a Join the boxes below to show how the model helps to explain the energy changes in an electrical circuit. The first one has been done for you.

The battery pushes on the charge, raising its potential energy.	The water loses potential energy as it falls into the tray.
The resistor heats up as the moving charge does work on its atoms.	The pump pushes on the water, raising it up to where it has more potential energy.
The charge loses potential energy as it does work in the resistor.	The water heats up slightly when it splashes into the tray.
The potential energy lost by the charge in the resistor is the same as the potential energy it gained in the battery.	The water loses the same potential energy when it falls into the tray as it gained in the pump.

b As charge flows around a circuit, its potential energy changes. The charge gains potential energy in a battery and loses potential energy in resistors (or other components).

Fill in the blanks to complete the sentences below.

➨ The p_____ d_____ between two points in a circuit is measured in v_____.

➨ Voltage is another word for p_____ d_____.

➨ The bigger the p_____ d_____ between two points in a circuit, the more

energy is transferred between these points.

c Look at the circuits on the right. The batteries and lamps are identical. The only difference is that circuit B has two of the batteries in parallel.

Look at the comparisons in the table below. Put a tick (✓) in the column that correctly describes how they compare in the two circuits.

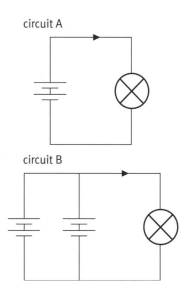

circuit A

circuit B

	the same	bigger in A	bigger in B
the potential difference across the bulb			
the current through the bulb			
the current through a single battery			
the time for the batteries to go flat			

d Potential differences add up around a circuit.

Look at the circuit on the right and the statements below. Draw a line to match each of the statements on the left with its explanation on the right. The first one has been done for you.

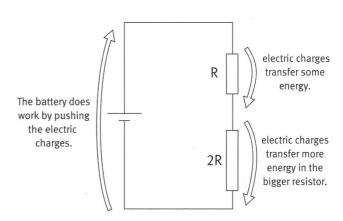

The battery does work by pushing the electric charges.

R

2R

electric charges transfer some energy.

electric charges transfer more energy in the bigger resistor.

The bigger the potential difference across the battery, the bigger its push.	Energy is conserved in the circuit: there is no net gain or loss of energy.
There is a drop in potential across each of the resistors.	The harder the battery pushes the charge, the more work it does.
There is a bigger potential difference across the bigger resistor.	The charge does work as it moves through a resistor.
The sum of potential differences across two resistors equals the potential difference across the battery.	The charge does more work as it goes through a bigger resistor.

8 Current and p.d. in circuits (series circuits)

a Complete the sentence below. Draw a ring round the correct bold words.
When two resistors are in series

➜ the current through the bigger resistor will be **bigger than / the same as / smaller than** the current in the smaller resistor.

➜ The potential difference across the bigger resistor will be **bigger than / the same as / smaller than** the potential difference across the smaller resistor.

b Look at the circuits below.

i For each circuit, put a tick (✓) next to the resistor which has the bigger potential difference across it.

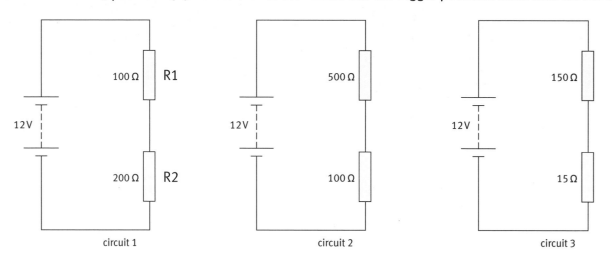

| circuit 1 | circuit 2 | circuit 3 |

ii In circuit 1, the total resistance is 300 Ω.

Calculate the current in the circuit. Use the equation current = voltage/resistance.

$$\text{current } I \quad = \quad \frac{V}{R} \quad = \quad \frac{\text{.............. V}}{\text{.............. } \Omega} \quad = \quad \text{.............. A}$$

iii Calculate the potential difference across each of the resistors in circuit 1.
Use the equation voltage = current × resistance.

p.d. across across R1 = ... A × .... Ω = ... V

p.d. across across R2 = ... A × .... Ω = ... V

iv Does this agree with the tick you put on circuit 1 in part **i**? ....

v Explain why the sum of the potential differences in part **iii** must be 12 V.

....

vi Draw in a voltmeter in circuit 3 to show how you would measure the voltage across the 150 Ω resistor.

9   Current and p.d. in circuits (parallel circuits)

a Complete the statements below. Draw a ring round the correct **bold** words.
When two components are in parallel

➔ the potential differences (voltage) across each one are **the same / different**
➔ the potential difference across each one is **equal to / smaller than** the voltage of the battery

b A hair dryer has a heater and a fan motor in parallel. There are two
switches. When switch **A** is closed, the fan comes on. When switch **B** is
closed, the heater comes on as well. The heater will not come on unless
switch **A** is already closed.

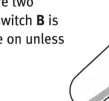

i Complete the circuit diagram below right by
labelling the two switches **A** and **B**.

ii Suggest why it is important that the hair dryer
cannot be turned on with the heater on and the fan off.

....

....

....

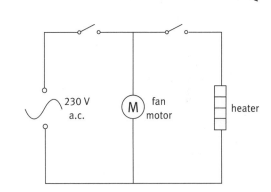

iii The motor has a resistance of 460 Ω and the heater
has a resistance of 46 Ω. The mains supply is 230 V.
Calculate

➔ the current in the 46 Ω heater

$$\text{current } I \quad — = \frac{V}{R} = \frac{\text{.............} V}{\text{.............} \Omega} = \text{.............} A$$

➔ the current in the 460 Ω motor

$$I = \frac{V}{R} = \text{.............} A$$

➔ the total current from the 230 V supply = ... A

iv Complete these sentences. Draw a ring round the correct **bold** words.

When two components are in parallel, the current in each one is **the same as / smaller than** the current
for that component on its own. The total **current / voltage** taken from the power supply will be
more / less than it was with just one component on its own. This extra current will have a cost: the time
for the batteries to go flat will **increase / stay the same / decrease**.

v Look at the list of words below. Draw a line under those that go *through* a component and a ring round
those that go *across* a component.

potential difference	flow	current	voltage

10 Power, fuses and cables

a In electric circuits, the equation for power is

power	=	potential difference (voltage)	×	current
(........................)		(........................)		(........................)

i Complete the equation above by putting in the units.

In symbols this is $P = VI$.

ii Rearrange the equation so that the current, I, is the subject:

$I =$ ☐

iii A 60 W bulb is connected to the mains. The mains voltage is 230 V. What current will it take? (You may find a calculator useful here.)

$I =$ ☐ $=$ .... A

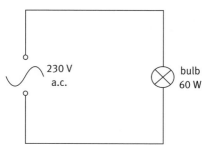

b Power measures the rate at which energy is transferred.
Complete the sentences below. Draw a ⟨ring⟩round the correct **bold** words.

A 100 W light bulb is **brighter / dimmer** than a 60 W bulb. It transfers energy **more / less** quickly.
They are both connected to the same, mains voltage of **230 V / 12 V**. This means that **more / less** charge
has to flow through the 100 W bulb each second. So the current in the 100 W bulb is **bigger / smaller**.

c In a house, light bulbs are connected in parallel.
The picture on the right shows a circuit with three 100 W bulbs.
Each bulb takes a current of 0.44 A.

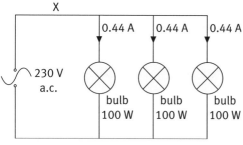

i What will the current be at the point marked X? .... A

ii The cable in the circuit has a current rating of 5 A.
To be safe, the current in these cables must be less than 5 A.
How many 100 W bulbs can be put in parallel before the
current in the cable becomes unsafe at the point marked X?

....

iii Why is it dangerous for the current in the cable to go over 5 A? ....

....

iv A 5A fuse is put in a house lighting circuit to protect the cables. On the circuit above, mark the place to
put the fuse.

v Explain how the fuse protects the cables in this circuit. ....

....

11 Domestic appliances and bills

a Power measures how fast we transfer energy, so

energy transferred = power × time

i complete these sentences.

➡ The equation 'energy = power × time' gives the energy in joules when power is measured in

.... and time is measured in ....

➡ A joule is a small unit for everyday purposes, so when we pay electricity bills the unit of energy is the

kilowatt hour (kWh).

➡ To calculate the energy in kWh, we use the same equation 'energy = power × time' but we measure

power in .... and time in ....

ii Asif puts a 60 W light on for 3 hours.

➡ Write 60 W in kilowatts: .... kW
➡ How many units (kWh) does the light bulb use in that time?

Energy = power × time

= .... kW × .... hour = ... kWh

iii Electricity supply companies charge about 10 p for each kWh. How much does it cost to have the light on?

cost = 10 p per kWh × .... kWh = ... p

b Filament light bulbs are not very efficient. They produce light by getting very hot. The diagram below
shows how the energy supplied to a light bulb by electricity is then transferred away from the bulb.

i How does the diagram show that only 25% of the energy supplied is usefully transferred?

....

....

....

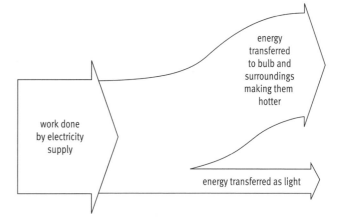

energy transferred to bulb and surroundings making them hotter

work done by electricity supply

energy transferred as light

ii Name a kind of light bulb that is more efficient.

....

c The equation for efficiency is:

$$\text{efficiency} = \frac{\text{useful energy transferred}}{\text{total energy supplied}} \times 100$$

An electric saw does 150 J of useful work for every 400 J supplied by the mains. What is its efficiency?

Efficiency   = ...   = ... %

12 Generator effect

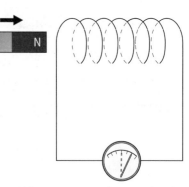

a A moving magnet induces a voltage in a coil of wire.
The picture shows the north pole of a magnet moving into a coil.
The needle on the ammeter flicks to the right.

Look at the boxes below. Draw a line to match each of the actions on the
left with one of the movements of the needle on the right.
You can use each needle movement once, more than once, or not at all.

This ammeter reads zero when its needle is in the middle.

Pull the north pole out of the coil

Hold the magnet stationary in the coil

Push the south pole into the coil

Pull the south pole out of the coil

flick to the left

no movement

flicks to the right

b The magnet can be put on a spindle and rotated near the coil. This will induce an alternating current in the coil.

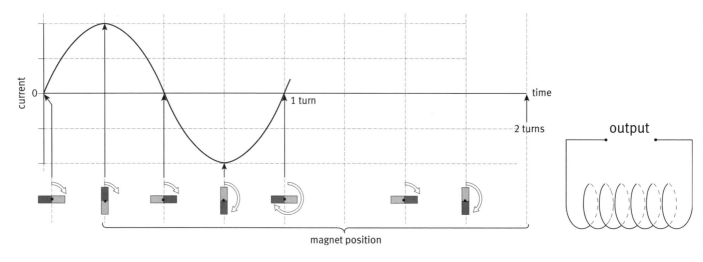

This diagram shows how the induced current varies as the magnet rotates through different positions. It shows one cycle of the a.c.

i Complete the diagram, showing how the current varies through two full rotations of the coil.

ii Draw in the missing magnet positions.

iii Complete the sentences.

In the generators of UK power stations, the time for one rotation is $\frac{1}{50}$ of a second. This means that

the frequency (number of cycles in one second) is .... hertz (Hz).

iv Write down four ways in which you could induce a bigger voltage in the coil.

....

....

13 Transformers

a Look at the picture of a simple transformer.

i Complete the diagram using these labels.

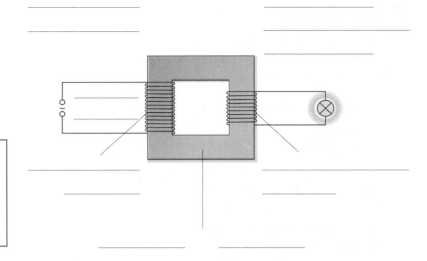

soft iron core	primary coil
secondary coil	a.c. supply
induced alternating current	

ii Complete these sentences. Draw a ring round the correct **bold** words.

A transformer works because the current in the **primary / secondary** coil produces a **magnetic / electric** field which passes through the secondary coil. This field is changing and therefore **induces / reduces** a voltage.

The transformer above has **more / fewer** turns on the secondary coil. This means it is a **step up / step down** transformer. The output voltage will be **more / less** than the input voltage.

iii This is the transformer equation: $\dfrac{V_p}{V_s} = \dfrac{N_p}{N_s}$

The input voltage is 230 V, and the number of turns are 460 (primary) and 24 (secondary). What is the output voltage from the transformer?

$$V_s = \text{_____} \text{ V}$$

iv The National Grid distributes mains electricity from power stations to people's homes.

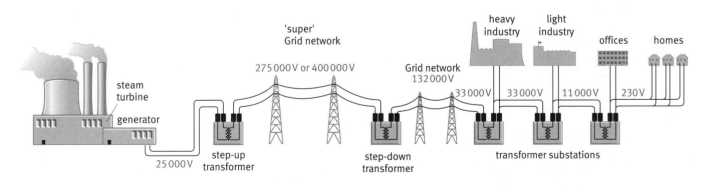

Look at the phrases below. Some of them apply to a.c. and some to d.c. Put a <u>straight line</u> under the phrases that refer to d.c. and a ͜curvy line under the phrases that refer to a.c.

It is easier to generate.   Its voltage has a constant value.   It can be distributed more efficiently.

It is produced by batteries.   It comes from the mains supply.   It won't pass through a transformer.

The wave model of radiation – Higher

1 Vibrations make waves

a The table shows some examples of waves.

In each case:

➔ underline the medium that carries the wave (the first one has been done for you)

➔ suggest what might have started the wave

Wave	What might have started the wave?
a sound wave in the <u>air</u>	
a ripple on the water of a pond	
a pulse on a slinky	
a wave on a stretched elastic rope	

b Complete these sentences. Draw a ring round the correct **bold** words.

The coils of a slinky spring are **disturbed / increased** by a wave. However, once the **disturbance / dismay**

has passed, the coils return to **where they started / the end of the spring**.

c Look at the quantities below. Draw a ring round those that can be permanently displaced by a wave – the wave carries them from one place to another.

energy	matter	particles	information

d i Use these words to complete the boxes on the left.

transverse   **longitudinal**

direction of wave

.... waves – when the particles of the medium vibrate at right angles to the direction in which the wave is moving

.... waves – when the particles of the medium vibrate in the same direction as the wave is moving

direction of wave

The pictures on the right show the two different types of wave on a slinky spring. The waves are moving left to right.

ii Draw arrows on each wave to show the direction in which the coils are moving.

iii Draw lines to link each box on the left with one of the waves on the right.

e Use these words to complete the sentences.

second	pitch	vibrating	hertz

Sound is produced by something _____ . The frequency of the vibrations determines the

_____ of the sound. The unit of frequency is the _____ (Hz), which measures

the number of vibrations every _____ .

f i Draw a line to match each note on the left with the type of vibration that produces it.

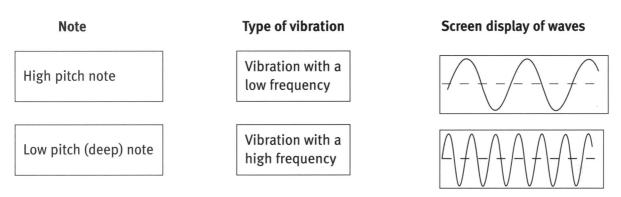

| Note | Type of vibration | Screen display of waves |

The sound waves can be picked up by a microphone and displayed on a screen. These are shown on the right.

ii Draw a line to match each type of vibration with one of the screen displays on the right.

g Cross out the wrong definitions to complete this table for a transverse wave.

amplitude	The maximum distance each particle moves from its normal position	or	The total distance from the top of a crest to the bottom of a trough
frequency	The time it takes a wave to pass a point	or	The number of waves passing a point every second
wavespeed	The speed at which the wave moves up and down	or	The speed at which the wave moves through the medium
wavelength	The length of a complete wave, e.g. from one crest to the next	or	The total distance that a wave has travelled from its source

h Mark on this diagram of a transverse wave:

➔ the **amplitude**

➔ the **wavelength**

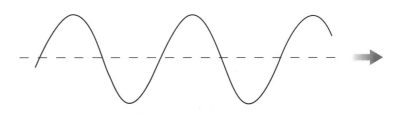

2   How waves move

The frequency of a wave depends on the source – how many times it vibrates per second.
The wavespeed depends on the medium the wave travels through.

a Wavespeed can be calculated using this equation:

wavespeed   =   frequency   ×   wavelength

(...)   (...............................)   (...............................)

  i Complete the equation by putting in the units.

  ii Rewrite the equation using these symbols: λ f v .... = ... × ....

  iii What is the symbol for wavelength? ....

b The diagram shows a burst of three
waves on a rope.

The frequency is 3 Hz and their
wavelength is 4 m.

Write your answers on the diagram.

  i How many complete waves pass a
point in 1 second?

  ii What is the wavelength of each wave?

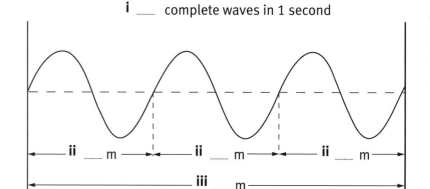

  iii What is the total length of this burst of three waves?

These three waves pass a point in 1 second.

  iv Complete this equation to find the speed of the wave:

  wavespeed = 3Hz × 4 m = ... m/s

c Another wave has a wavelength of 2 m and a frequency of 6 Hz.
Use the equation to find the speed of this wave.

wavespeed = ... × .... = ... m/s

d The waves in parts **b** and **c** were made on the same piece of rope.

Complete these sentences. Draw a (ring) round the correct **bold** words.

The speeds of the waves in parts **b** and **c** are **the same / different**. This is usually the case for waves in the

same **medium / direction**: their speed **does / does not** change even if they have different frequencies.

In summary: usually the speed of a wave is **independent of / proportional to** its frequency.

3   Water waves

a The diagrams below show four ways in which water waves behave. Use the words in the list to label the diagrams.

refraction	reflection	interference	diffraction

i

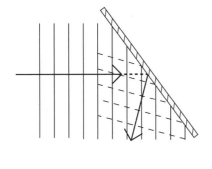

....

ii

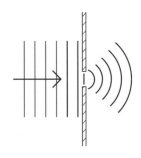

....

iii

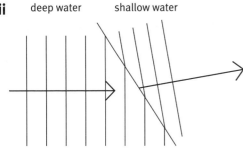

....

iv

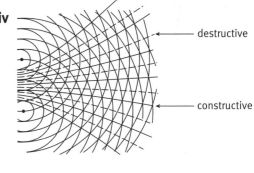

....

b The picture below right shows a water wave going from deep water to shallower water.

Choose words from this list to complete the sentences.

shallow	deep	frequency
wavelength	shorter	longer
slowed	reduced	

Water waves travel faster in .... water than in .... water. As the wave

crosses the boundary, the .... of the wave stays the same but the wavelength gets

.... This is because the waves are .... down; the progress of each wave is

.... by the time the next wave arrives.

c The wave in the picture is refracted.

   i Put a tick (✓) in the region where the waves have a shorter wavelength.

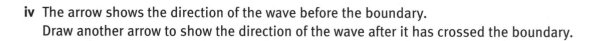

   ii Are the waves slower or faster in this region? ....

   iii Label the two regions 'slow' and 'fast' to describe the speed of the waves.

   iv The arrow shows the direction of the wave before the boundary.
Draw another arrow to show the direction of the wave after it has crossed the boundary.

   v The statements below explain why its direction changes. They are out of sequence. Use arrows to join the boxes in the correct order. The first one has been done for you.

reaches the shallow water first. It		stays at the same speed. The wave

is slowed down while the rest of the wave

The top part of the wave		is skewed upwards at this boundary.

d Waves will spread out when they pass through a gap in a barrier.

wave approaching barrier   wave after it has passed through

   i What is the name of this effect?

   ....

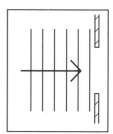

   ii The picture on the right shows some waves approaching barriers and the waves after they have passed through the barrier.

   Draw lines to match each barrier with the wave after it has passed through.

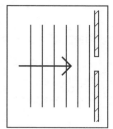

   iii What happens to the spread of the wave as the gap gets wider?

   ....

   ....

e Complete these sentences. Draw a ⟮ring⟯ round the correct **bold** words.

Diffraction is most noticeable when the width of the slit is about the same size as the

wavelength / amplitude of the wave. Light has a very **long / short** wavelength. A light wave will diffract

when it passes through a very **narrow / wide** slit.

4   Sound and light – interference

a Interference is a wave effect.

⮑ Link up all the boxes below that describe **constructive interference**. Try to do them in a logical order. The first one has been done for you.

⮑ Now link up all the boxes that describe **destructive interference**. Try to do them in a logical order. Use a different colour.

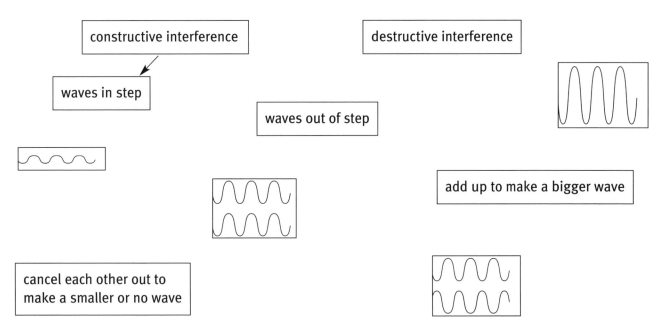

b The picture shows two loudspeakers connected to a signal generator. Each speaker is producing the same pure tone. The curved lines show a snapshot of the peaks of the sound waves.

i Fraser is standing at the point marked F.
Complete these sentences. Draw a ⟨ring⟩ round the correct **bold** words.

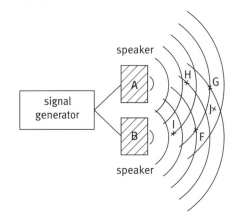

When a peak from loudspeaker A reaches the point F,

a **peak / trough** from loudspeaker B reaches the same point.

The waves will **add up / cancel out**. Fraser will hear a sound that

is **louder / quieter** than one of the loudspeakers on its own.

ii There are people standing at the points G to J. Draw a circle round each person who is at a point of constructive interference.

iii Interference is a feature of waves. Does this experiment suggest that sound is a wave?

c We can also get interference effects with light.
Draw a ⟨ring⟩ round the piece of apparatus that is often used to produce an interference pattern.

telescope	mirror	double slits	prism	lens

5 Is light a wave?

a Put a tick (✓) in the correct columns to show which behaviour is good evidence for sound and light being waves.

Behaviour of light	This happens to waves	Also happens to particles
reflected from a mirror		
diffracted through tiny gaps		
refracted as it passes from air to glass		
interference when two narrow slits produce bright and dark fringes		

Behaviour of sound	This happens to waves	Also happens to particles
produces echoes		
diffracted through an open door		
refracted between hot and cold air		
interference when two speakers produce loud and quiet regions		

b The picture shows a light ray at a boundary between air and water.

   i Light travels more slowly in water. Label the two sides of the boundary to show which side is air and which side is water.

   ii This ray of light is travelling from air to water. Add an arrow to the ray to show its direction.

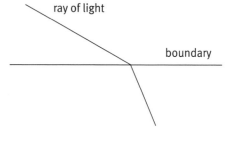

c The pictures below right show two rays of light (A and B) meeting the boundary between glass and air.

   i Complete the sentences. Draw a (ring) round the correct **bold** words and fill in the blanks.

      Ray A is **refracted / diffracted** as it passes through the boundary

      between glass and air. The **angle / intensity** of this ray means that

      it only just gets out of the glass. Ray B strikes the boundary at a

      shallower / steeper angle. It **can / cannot** get out of the glass.

      Instead, it is **reflected / absorbed** back into the glass. This is called

      T .... I .... R ....

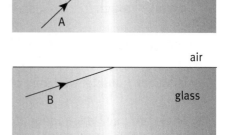

   ii Draw in the path of ray B after it has been reflected.

d The picture shows an optical fibre with a ray passing in at the left-hand end.

   i Draw the path of this ray as it goes through the fibre.

   ii Give two uses of optical fibres.

      ➔ ....

      ➔ ....

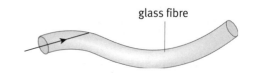

6 Electromagnetic waves

a Complete the sentences. Fill in the gaps and draw a (ring) round the correct bold words.

Light is part of a larger spectrum of **waves / sounds** called the E _____ S _____ .

All of these waves travel at **the same / their own** speed in a vacuum. It is exactly **300 000 / 0.1** km/s. The

shortest waves are **gamma rays / X-rays** whose wavelengths are around a billionth of a millimetre (about

the size of a nucleus in **an atom / a plant cell**). The longest waves are **radio / ultraviolet** waves whose

wavelengths are a few kilometres (the size of a small **town / person**).

b The picture shows the electromagnetic spectrum.

i Fill in the boxes with the names of the regions. Visible light has been done for you.

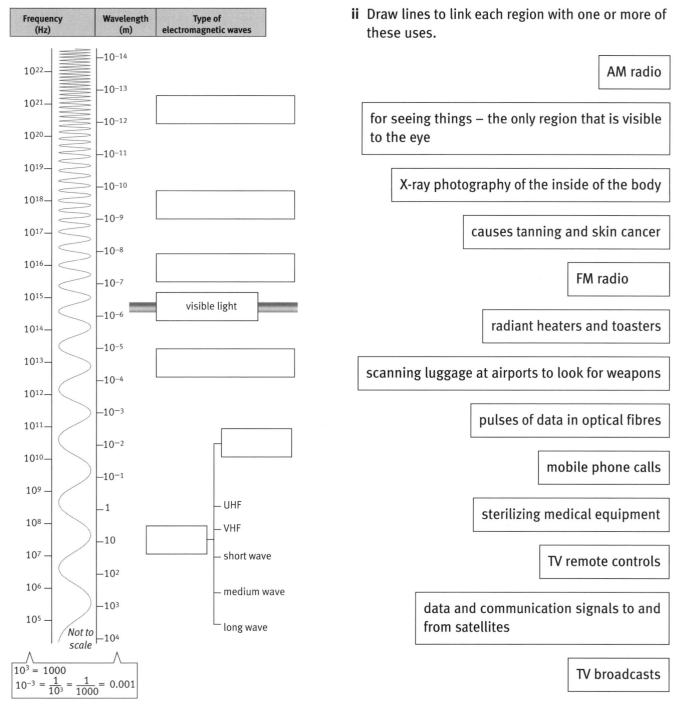

7 The risky side of the rainbow

a There are three types of radiation with a wavelength that is shorter than visible light. The table represents this part of the electromagnetic spectrum.

Part of the electromagnetic spectrum
visible light

  i Fill in the names of the three types of radiation. Put the one with the shortest wavelength at the top.

  ii Is the frequency of this radiation higher or lower than the frequency of

  visible light? _____

  iii The radiation is carried as photons. Do these photons have more or less energy than the photons of visible light?

  iv Put a tick (✓) next to the part of the spectrum where a photon has the most energy.

b Choose words from this list to complete the sentences.

cells	ionizing	high	low	electrons
molecules	randomizing	ionize		photons

When electromagnetic waves with a _____ frequency strike atoms, they can _____

the atoms. Their _____ carry enough energy to knock out _____ . This can alter

the material. If this happens in living _____ , they can be damaged or destroyed. You should take

precautions to reduce your exposure to these _____ radiations.

c Val puts on sunblock when she goes on the beach on hot days.

Complete these sentences. Draw a ⟨ring⟩ round the correct **bold** words.

The sunblock protects Val by **absorbing / transmitting** some of the ultraviolet radiation. Some radiation

still reaches her skin. It has the same **wavelength / intensity** as before but a lower **wavelength / intensity**.

So its photons have the **same / more** energy and the number getting through has been **reduced / unaffected**.

This **reduces / increases** the risk of damage to her skin cells.

d A radiologist uses X-rays to look for broken bones. Explain why

  i the radiologist stands behind a protective screen _____

  _____

  ii the radiologist wears a badge that measures the radiation dose _____

  _____

8 Infrared

a The picture shows white light from a filament lamp passing through a prism.

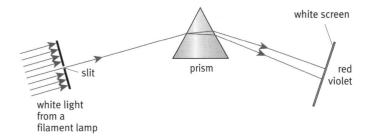

i Complete these sentences. Draw a ring round the correct **bold** words.

The light is **dispersed / combined** in the prism. This produces a **spectrum / reflection** of colours.

ii The speed of light in air is the same for all the colours. However, dispersion shows that the speed is

different for different colours in glass. Which colour is the slowest in glass? _____

iii An infrared detector is put in front of the screen. Put an X on the diagram to show where it would detect radiation.

b i Now draw a line to match each type of object (left) with a description of the radiation it emits (middle). One has been done for you.

ii Now draw lines to link each description to two examples.

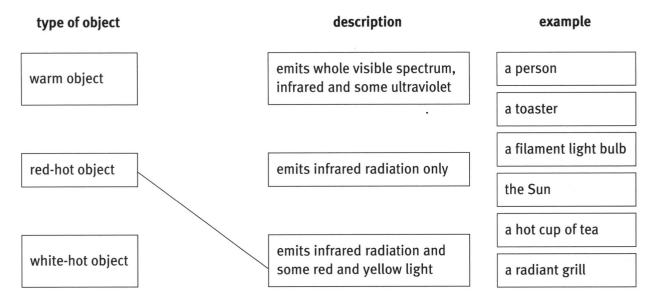

type of object	description	example
warm object	emits whole visible spectrum, infrared and some ultraviolet	a person
		a toaster
		a filament light bulb
red-hot object	emits infrared radiation only	the Sun
		a hot cup of tea
white-hot object	emits infrared radiation and some red and yellow light	a radiant grill

c Solve the clues to fill in the missing words about infrared (IR) radiation.

1 The type of radio wave that is next to infrared in the spectrum

2 Optical fibres allow infrared to pass through. They are . . .

3 This is lower for infrared than for visible light.

4 A type of imaging that can find warm bodies in the dark or under rubble

5 What infrared radiation will do to your skin when you absorb it

6 A _____ control; it uses an infrared beam to change TV channels

7 How the wavelength of infrared compares to visible light

8 The colour in the visible spectrum that is next to infrared

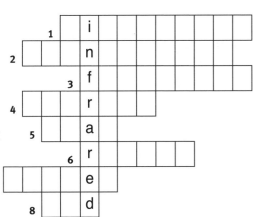

9   Microwaves, radio waves, and SETI

a Look at the uses of microwaves below. Draw a line to match each use with a specific property of microwaves. The first one has been done for you.

carrying long distance telephone calls	They pass through the atmosphere.
cooking food	They can be beamed in a straight line.
carrying satellite TV signals	Some wavelengths are absorbed by water molecules.
carrying mobile phone signals	They can be produced by small electronic circuits.

b Look at the diagram. It shows the percentage of each type of radiation that reaches the ground from outside the Earth's atmosphere.

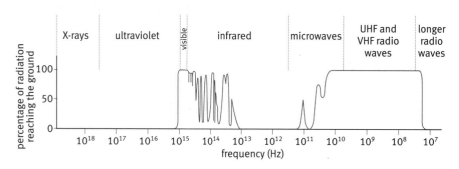

i   Put an 'A' under the types of radiation that are absorbed by the atmosphere.

ii Put a 'T' under the frequencies that are transmitted by the atmosphere.

iii Suggest two types of scientist who make use of the microwave 'window' in the atmosphere.

➜ ....   ➜ ....

c Look at the statements on the left. Each one applies to either sound waves or electromagnetic waves. Draw a line to link each one with the type of wave it describes. The first one has been done for you.

They are longitudinal waves.	
They travel through a vacuum.	
Their speed is about 300 000 km/s in air.	sound wave
Their speed is about 300 m/s in air.	
They need some matter (solid, liquid, or gas) to carry them.	electromagnetic wave
They are transverse waves.	

10   Getting waves to carry information: AM

Waves can transmit information (a signal) if the amplitude of the wave is changed to match the signal.

a Draw a line to match each description with the correct wave.

audio frequency wave (AF)	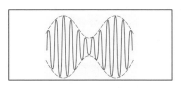
amplitude modulated wave (AM)	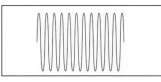
radio frequency carrier wave (RF)	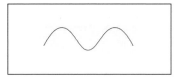

b Complete the sentences below. Draw a (ring) round the correct **bold** words.

Most humans can hear sounds up to a **frequency / wavelength** of about 20 kHz. A radio carrier wave has to

have a frequency that is much **greater / smaller** than this. This means that there is a number of complete

cycles of **carrier / audio** wave for each cycle of the **carrier / audio** wave.

c In the UK, there are dozens of radio stations that broadcast in the Medium Wave band, which covers frequencies from 620 kHz to 1630 kHz. Each radio station has its own broadcast frequency. Explain why

i   each station needs its own frequency: ....

ii a radio receiver needs a tuning circuit: ....

d The list on the left shows some effects on AM broadcasts in the Long Wave band. The causes are shown on the right. Draw a line to link each effect to its cause.

The signal can pick up stray pops and buzzes from electrical equipment nearby.	The wave has a long wavelength which diffracts around geographical features.
The signal strength varies on motorways.	The wave is amplitude modulated so any interference that affects the amplitude becomes part of the signal.
The radio stations can be received behind hills and homes.	Waves from two transmitters can interfere with each other and produce peaks and troughs in the signal strength.

c The diagram shows a simple radio transmitter (on the left) and receiver (on the right).

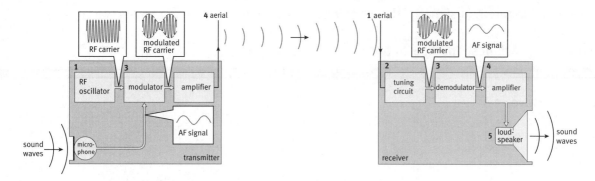

Note: you do not need to remember the details of these diagrams. They are to help you understand how the modulated wave is made and what it looks like.

i The boxes below describe the stages in producing and broadcasting an AM signal. They are out of sequence. Draw arrows to join the boxes in the correct sequence. The first one has been done for you.

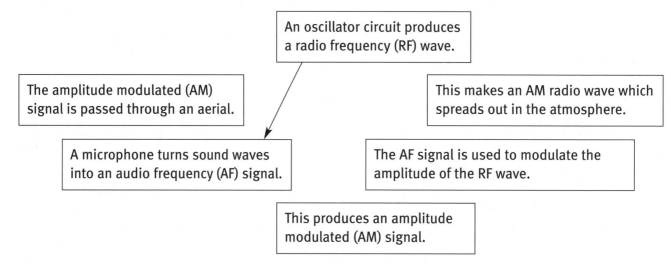

ii The radio receiver picks up the AM radio wave and produces a sound wave. The list on the left shows some of the parts of a radio receiver. They are described on the right. Draw a line to match each part with its description. The first one has been done for you.

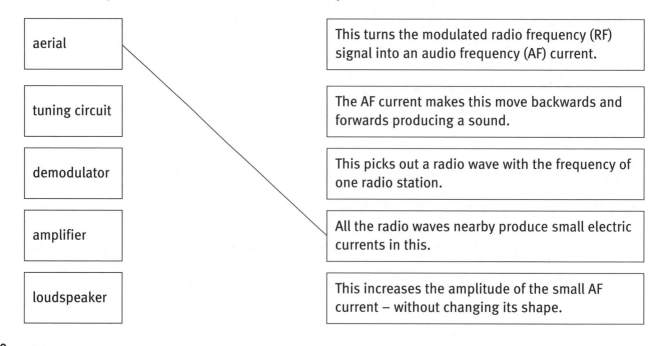

11 FM and digital

a Use these words to complete the descriptions below. (You need to use some words more than once.)

> audio frequency (AF)   frequency   amplitude modulated (AM)
>   amplitude   frequency modulated (FM)

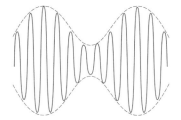

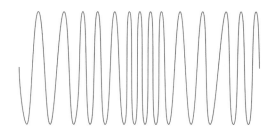

This signal is ....

The changes in the ....

exactly match the changes in the original

.... signal.

This signal is ....

The changes in the ....

exactly match the changes of the original

.... signal.

b Both the signals in the previous question are analogue signals. They change continuously. A digital signal does not vary continuously. The voltage is sampled many times a second and each value is given a code. The codes are **binary**. All the numbers contain only two digits – 0 (no pulse) and 1 (a pulse).

The diagram shows how an analogue signal can be sampled to give a digital signal.

Follow these steps to complete the diagram.

In the boxes:

➔ write in the value of the voltage – to the nearest whole number
➔ write in the binary number that has the value in the box above
➔ draw the pulses to represent the binary number

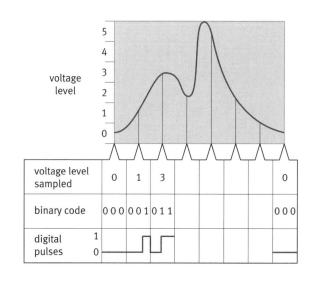

c complete the sentences below. Draw a (ring) round the correct bold words.

Digital signals can be sent along optical **strings / fibres**. The 1s and 0s are used to switch a **laser / lamp**

on and off; it **emits / absorbs** millions of **infrared / ultraviolet** pulses every second. Fibres can be many

kilometres / millimetres long. They are made from special **glass / steel** that is extremely pure so that

hardly any of the radiation is **absorbed / transmitted**.

12 Going digital

a Digital signals are replacing many analogue signals. But analogue signals are still used.

The table shows some technologies which have analogue and digital versions.

⮞ Fill in the **Digital** and **Analogue** columns to show the media and methods used in each version.

Use the words and abbreviations in this list. Each word is used once only.

DAB	35 mm film	Hard disk/MP3	AM	Cassette	CCD

Technology	Digital	Analogue	Advantages of digital
cameras			
talk radio			
portable music player			

⮞ For each technology, write in an advantage of using digital signals.

b The diagram below shows the transmission of an analogue signal and a digital signal.

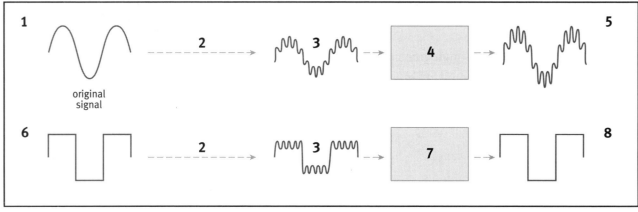

i Match these labels with the numbers on the diagram. Write the correct number in each box.

☐ weak signal with noise   ☐ regenerated signal   ☐ amplifier   ☐ digital signal

☐ amplified signal with noise   ☐ analogue signal   ☐ transmission   ☐ regenerator

ii Which word can be defined as 'low level, unwanted waves that contaminate the original signal'?

N..

iii Which type of signal can produce a perfect copy of itself at the destination?

iv Explain why this is such an advantage. ....

....

This page is blank

1 Observatories and telescopes

An observatory is a site used for observing objects in space. A telescope is any instrument that collects radiation from these objects for astronomers to study.

Complete the table, giving examples of telescopes and their locations.

Telescope	Type of electromagnetic radiation detected	Location
the Monument		
		Germany
	radio	
		low Earth orbit
Calar Alto		
	visible light	

2 How telescopes work

Telescopes can make things visible that cannot be seen with the naked eye.

Describe two different ways that they might do this. Use words from the box in your description.

distant source	telescope	weak radiation
detector	part of the electromagnetic spectrum	

....

....

....

....

3 Computer control in observatories

Tick the phrases below that complete this statement correctly. Tick more than one.

Computer control enables a telescope to

track a distant source, collecting weak radiation from it, while the Earth rotates ☐

scan a distant source, collecting data from each part of it ☐

see clearly even when viewing conditions are poor ☐

follow instructions from an astronomer who is not based at the observatory ☐

move quickly and point precisely to another part of the sky ☐

operate for more hours each year of its lifetime ☐

4 Converging lens

Complete the diagram to show what happens to the rays after passing through the lens.
Label the principal axis, focus, and focal length of the lens.

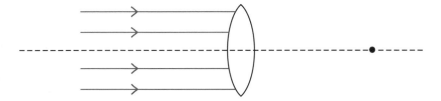

5 Converging lens shapes

Describe how the shape of a stronger lens differs from the shape of a weaker lens.

6 Power of a lens

a If you know the focal length of a lens, you can calculate its power. Give the equation for this.

b Calculate the power of a lens with each of these focal lengths:

i 0.5 m

ii 40 cm

iii 20 cm

iv 5.0 m

7 Describing images

Complete this statement using words from the box.

inverted	same size	real	bigger

What kind of image a converging lens makes depends on how far away an object is from the lens. An image

can be described by its size compared with the object: smaller, the _____ _____

as or _____

orientation compared with the object: upright or _____

image type: virtual or _____

8 Lenses in a telescope

a Label the eyepiece and objective lens in the diagram of a telescope below.

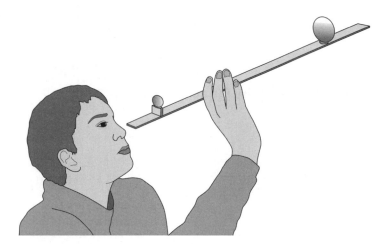

b Describe how the shapes of the two lenses compare.

_____

_____

9 Converging and diverging lenses

In these diagrams, label the converging lenses **C** and the diverging lenses **D**.

10 The magnification of a telescope

a Explain what is meant by the magnification of a telescope.

....

....

b Although no telescope can make stars look any larger, a telescope with greater magnification is still better for observing stars. Explain why.

....

....

c The magnification of a telescope can be calculated using this relationship:

$$\text{magnification} = \frac{\text{focal length of objective lens}}{\text{focal length of eyepiece lens}}$$

Complete the table below by calculating the magnification of a refracting telescope with each pair of lenses.

Telescope	Focal length of objective lens	Focal length of eyepiece lens	Magnification
A	50 cm	5 cm	
B	80 cm	5 cm	
C	5.0 m	5 cm	
D	1000 mm	4 mm	
E	2000 mm	10 mm	

d Which telescope provides the greatest magnification? ....

---

11 The aperture of a telescope

The aperture of a telescope is the light-gathering area of its objective lens. Give one reason why a telescope with a larger aperture is better than one with a smaller aperture.

....

....

12 Comparing lenses

The table below lists lenses, all made from the same type of glass, which may be used in making a telescope.

Lens	Focal length (mm)	Lens diameter (mm)
P	500	80
Q	250	120
R	25	60
S	50	100

a Which lens is the most powerful? ....

b Which lens would be thinnest (have surfaces with the least curve)? ....

c Which lens, if used as the objective lens, would give the brightest image? ....

d Calculate the magnification of a telescope made using lenses P and R.

....

....

13 Reflecting telescopes

Referring to the diagram below, explain a major problem with refracting telescopes. Explain also how this problem is overcome by using a mirror as the telescope objective.

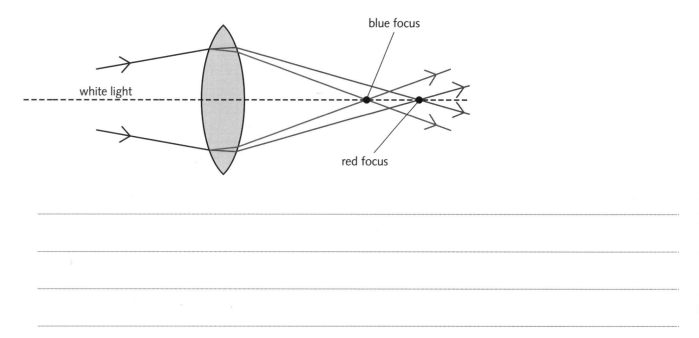

....

....

....

....

14 Mirrors for telescopes

Give three more reasons why most telescopes used by professional astronomers have mirrors as their objectives, and not converging lenses.

1 ....

....

2 ....

....

3 ....

....

15 Parabolic reflectors

A parabolic mirror is the most common shape for the objective of a reflecting telescope.

Complete the diagram to show what happens to the rays after striking the mirror. Label the principal axis, focus, and focal length of the mirror.

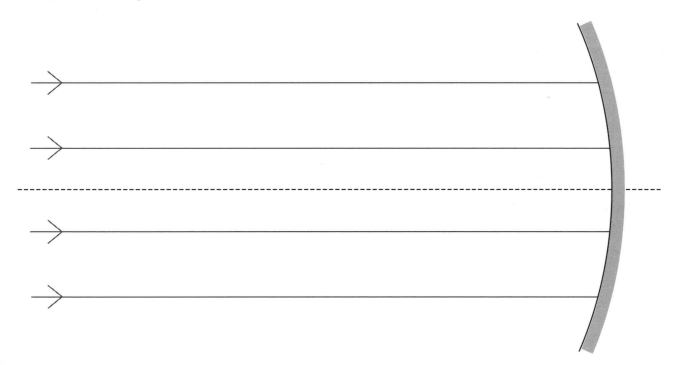

16 Wavelength and diffraction

Complete the diagram below to show what happens to the waves at these apertures.

17 Wavelength and resolving power

Use words from the box to complete these sentences.

resolving power	diffraction	separate	electromagnetic	aperture	wavelength

The _____ of a telescope is its ability to distinguish between

two closely spaced objects so that they are recognizable as _____ objects.

When the telescope _____ is large in relation to the _____ of the

_____ radiation being collected, astronomers will see more detail in an image. This is

because of the effect of wave _____ at any opening.

18 Images and resolving power

These diagrams show two light sources, observed through three different telescopes.

Draw lines to match each image to the telescope that made it.

High resolution telescope

Medium resolution telescope

Low resolution telescope

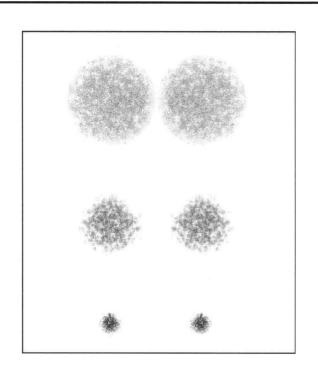

19 Large telescopes

Give two reasons why astronomers need to build very large telescopes.

1 ....

....

2 ....

....

20 Building telescopes

Describe two different engineering challenges in building large telescopes.

1 ....

....

2 ....

....

21 Parallel rays

Explain why light from any star arrives at a telescope as parallel rays. Use a diagram to help.

22 The image formed by a converging lens

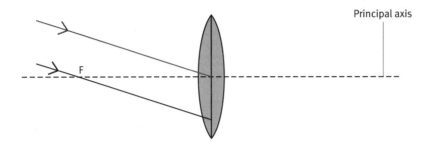

a The diagram above shows two rays from a distant star. One strikes the centre of the lens. The other passes through the lens focus before striking the lens. Show what happens to each ray, and add labels to describe the rule for each of them.

b Label the place where the lens produces an image of the star.

c Draw two more rays coming from the same star, and passing through the lens.

d Is the image virtual or real? Explain why.

23 An extended object

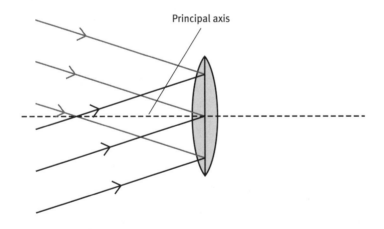

a The diagram above shows rays from a distant galaxy. The telescope objective lens gathers light from two sides of the galaxy. Assume the central rays both pass through the focus of the lens. Complete the rays to show where the lens produces an image of the galaxy.

b Annotate the diagram to explain the result.

c A galaxy is one example of an extended object that an astronomer might study. Give another example.

24 Refracting telescope objective

Explain why the objective of a refracting telescope should have a large diameter but be a weak lens.

25 The motion of stars

8 p.m.   midnight

☆ Pole Star   ☆ Pole Star

horizon looking north   horizon looking north

a Describe how the stars appear to rotate during this 4 hour period.

b Explain why the stars appear to rotate.

c Why does the Pole Star not appear to move?

26 Sidereal and solar days

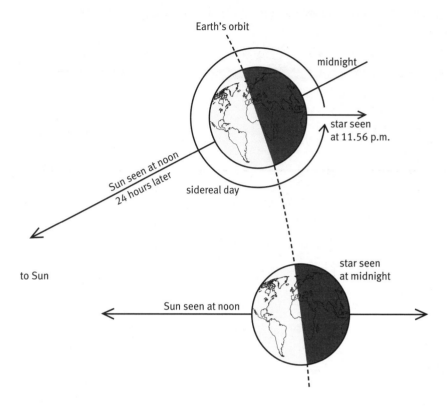

a Add labels to the diagram above, showing the time in hours and minutes
for a sidereal day
for a solar day

b Use the diagram to explain why any particular star rises 4 minutes earlier each day.

....

....

c Explain why a star will appear at exactly the same time and position in the night sky once every year.
(Show that 4 minutes later every day, for 365 days, is about 24 hours – one full rotation of the Earth.)

....

....

....

27 Sun and Moon

Both the Sun and the Moon rise in the east, and cross the sky before setting in the west.

This sequence of diagrams shows the Moon moving east to west across the sky, but slipping back though the pattern of stars.

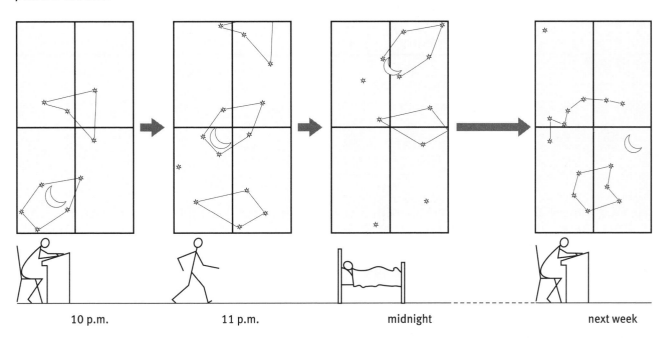

10 p.m.   11 p.m.   midnight   next week

a Fill in the gaps, giving the times:

The stars appear to travel east to west across the sky once every ....

The Moon appears to travel east to west across the sky once every ....

b Explain why this happens, mentioning the Earth's rotation and the Moon's orbit of the Earth.

....

....

....

....

....

....

28 Phases of the Moon

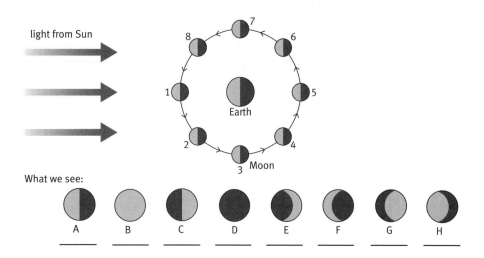

a Match diagrams A–H showing what we see with positions of the Moon 1–8 as it orbits the Earth.

b Write a short paragraph explaining why the appearance of the Moon changes in a regular way, in terms of the relative positions of the Sun, Moon, and Earth.

29 The position of a star

Draw a diagram and explain how you could use two angles to describe the precise position of an astronomical object.

30 Constellations and seasons

Different star constellations are visible in the night sky at different times of the year.

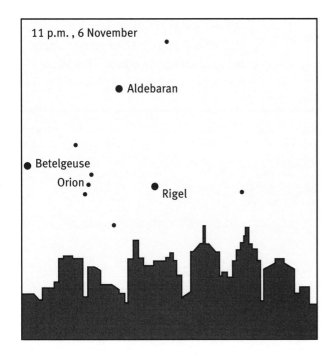

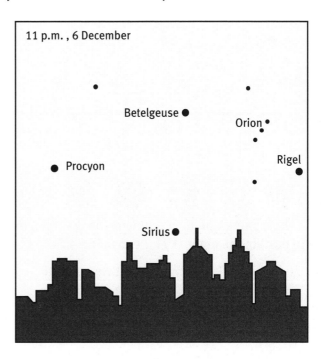

Explain this observation by annotating the diagram below, which shows the Earth orbiting the Sun and also shows the constellations of the zodiac.

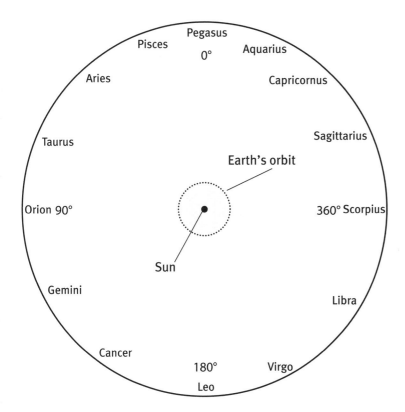

The zodiac is the set of constellations located around the ecliptic, i.e. around the extended plane of Earth's orbit of the Sun.

31 Retrograde motion of planets

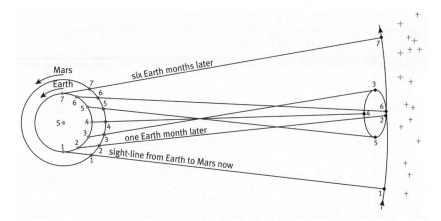

From months 1 to 3, Mars appears to move forwards. Then, for two months, it goes into reverse before moving forwards again. This is known as retrograde motion.

Write a short paragraph to accompany the diagram, explaining why Mars appears to move backwards (west to east) in the night sky for two months.

32 Eclipses

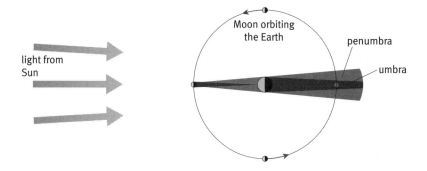

a Add labels to show the Moon's position when it
  i causes a solar eclipse
  ii causes a lunar eclipse

b Use the diagram to explain why lunar eclipses can be seen by many more people than solar eclipses.

....

....

....

33 Why eclipses are rare

Draw a diagram and annotate it to explain why eclipses are not seen every month.

34 Parallax

Using an everyday example, explain why nearby stars appear to move during the year against the background of distant stars.

35 Parallax angle

a Draw a diagram in the space below to define parallax angle.

b Use your diagram to explain why a star with a smaller parallax angle is further away.

36 Star distances

Complete the following sentences.

There are _____ ° (degrees) in a full circle.

There are _____ ' (minutes) of arc in 1o.

There are _____ " (seconds) of arc in 1'.

A unit of distance based on the measurement of parallax is the _____.

An astronomical object at a distance of 1 parsec has a parallax angle of _____. This distance

is equivalent to about _____ light-years.

37 Parallax and parsecs

Complete the table by calculating each distance in parsecs.

Parallax angle (seconds of arc)	Distance (parsecs)
0.769	
0.1	
0.025	
0.0125	
0.06	
0.01	

38 Luminosity

The rate at which a star radiates energy is called its luminosity. Underline two factors that can affect a star's luminosity:

size   twinkling   parallax angle   temperature

39 Observed brightness of a star

Explain why the observed brightness of a star (seen by an astronomer) will depend on both its distance and its luminosity. It may help to draw a diagram.

40 Comparing Betelguese and Orion

The constellation Orion contains the bright stars Betelgeuse and Rigel.

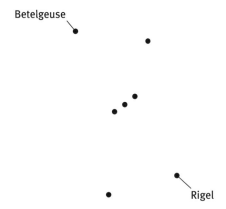

The following statements about these stars are all true.

Betelgeuse and Rigel are both about the same size.
Betelgeuse is red, while Rigel is blue-white.
Rigel gives out more than four times as much energy every second as Betelgeuse.
From Earth, Betelgeuse appears slightly brighter than Rigel.

Use this information to write a comparison of the nature of the two stars.

41 Star temperatures

Explain how astronomers analyse starlight to work out a star's temperature. Use the words in the box to help you.

electromagnetic radiation	frequencies	temperature	spectrometer
peak frequency	telescope	intensity of radiation at each frequency	

....

....

....

....

....

....

....

42 Star spectra and luminosity

The spectrum of a star can also be used to estimate its luminosity.

Tick the correct ending for the following statement.

Knowing the luminosity of a star and the intensity of its light at the telescope, an astronomer can calculate its

temperature ☐

distance ☐

colour ☐

galaxy ☐

43 The spectrum of a star

The graph shows the spectra produced by three different stars.

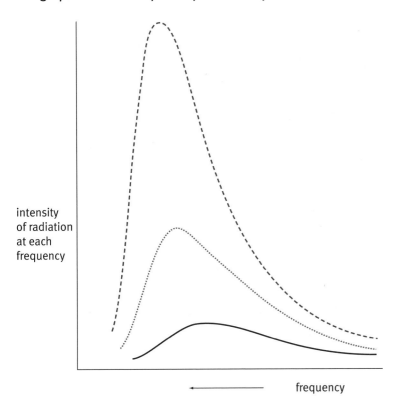

a Label the hottest star and the coolest star.

b What does the area under each graph indicate about the radiation from the star?

44 Cepheid variable stars

Cepheid variable stars were important in working out the distances to other galaxies.

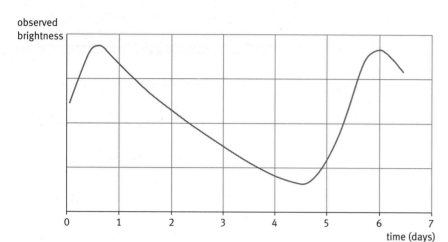

a The graph shows the brightness curve for a Cepheid variable star. Describe the pattern in its observed brightness.

....

b Henrietta Leavitt made a very important discovery about Cepheids. Draw a line linking the two variables which she found were related.

luminosity
temperature
distance

observed brightness
period
speed of recession

c The following sentences describe how astronomers use a Cepheid variable star to work out the distance to another galaxy. Number the boxes to put them in the correct order.

☐ Use the period of variation of the Cepheid variable star to estimate its luminosity.

☐ Produce a light curve for the Cepheid variable star and measure its period of variation.

☐ Knowing both the luminosity of the star and the intensity of its light at the telescope, calculate the distance to the Cepheid variable star.

☐ Look for a Cepheid variable star in the galaxy of interest.

5 Take the distance of the Cepheid variable star as the distance to the galaxy.

In practice, astronomers would calculate the distance to many Cepheid variables stars in a galaxy to estimate its distance.

45 The scale of the Universe

a Use words from the box to complete the sentences.

nebulae	Curtis	Universe	Milky Way	galaxies	Shapley

In 1920 there was a famous debate between two American astronomers, Harlow Shapley and Heber

Curtis, about the nature and size of the _____. Central to the debate was the interpretation

of thousands of fuzzy objects observed in the night sky, called _____. The Milky Way

includes nebulae and is much larger than previously thought, suggested _____. Spiral

nebulae are outside the _____, and are distant _____

similar to our own, suggested _____.

b How did Cepheid variable stars help to resolve this debate? Include the name of the astronomer responsible for this breakthrough.

....

....

....

46 Galaxies

The *Spitzer Space Telescope* is an orbiting infrared telescope. It was used to survey 30 million stars in the Milky Way, from which a group of astronomers was able to build up a picture of our galaxy published in 2005.

a What is a galaxy?

....

b Why was it essential to use an infrared telescope for this survey?

....

....

c Suggest why the survey was done by a *group* of astronomers.

....

....

d Complete the following sentences, using units in the box.

parsecs (pc)	kiloparsecs (kpc)	megaparsecs (Mpc)

Distances between stars in a galaxy are typically measured in _____ .

Distances between galaxies are typically measured in _____ .

47 Moving galaxies

Use words from the box to complete these sentences.

speed of recession	spectra	Hubble	away from	Cepheid variables

By analysing the _____ of _____ stars in 46 galaxies, _____

discover that all other galaxies are moving _____ ours. He also found that a galaxy's

_____ is proportional to its distance.

48 The Hubble equation

speed of recession = Hubble constant × distance

Complete the table showing data for different galaxies, using the Hubble equation.

Speed of recession	Hubble constant	Distance
5000 km/s	70 km/s per Mpc	_____ Mpc
3500 km/s	_____ km/s per Mpc	48 Mpc
_____ km/s	2.3×10^{-18} s^{-1}	3.08×10^{21} km
2000 km/s	2.3×10^{-18} s^{-1}	_____ km
3000 km/s	_____ s^{-1}	1.23×10^{21} km

49 The composition of stars

The diagram shows the dark lines seen in the spectrum of visible light from a star.

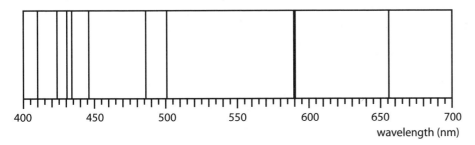

a Use the table below to identify the elements present in this star. Put a tick (✔) or a cross (✘) in every box in the last column.

Element	Wavelengths (nm)	Present in the star?
calcium	423, 431	
helium	447, 502, 588	
hydrogen	410, 434, 486, 656	
iron	431, 438, 467, 496, 527	
sodium	589, 590	

b A star is observed that contains only hydrogen and helium. Suggest and explain what this might imply about the star.

....

....

....

50 Energy levels and emission spectra

The diagram shows possible energy levels in an atom.

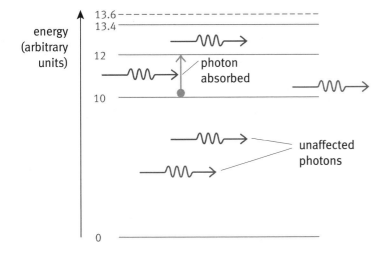

a Which subatomic particles are affected when atoms gain or lose energy by a discrete amount?

....

b Explain why this element emits light that makes a line spectrum rather than a continuous spectrum.

....

....

....

....

....

....

c The dotted line marked 13.6 represents the energy needed for ionization. An electron with this amount of energy, or more, would be able to escape from the atom. Use this mechanism to explain the difference between ionizing and non-ionizing radiations.

....

....

....

....

....

---

51 Atoms and nuclei

a The diagram below shows the apparatus used in Rutherford's alpha-scattering experiment.

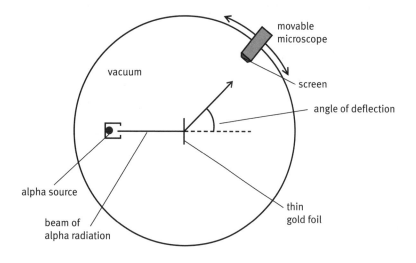

Why it is essential that the experiment is carried out in a vacuum chamber?

....

b Alpha radiation is directed at a sheet of gold foil. Describe the observations and conclusions from this experiment. (You may want to include a further diagram with the conclusions.)

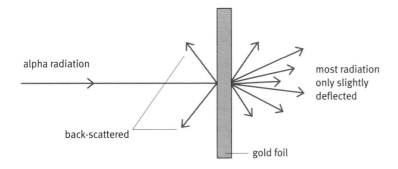

alpha radiation

most radiation only slightly deflected

back-scattered

gold foil

Observations:

Conclusions:

c Write these particles in the boxes below in order, from largest mass to smallest:

gold atom, alpha particle, gold nucleus, electron

largest

0.3 nm

smallest

52 The nucleus

a What two particles are found in the nuclei of atoms?

....

b Particles found in the nuclei of atoms are called nucleons. How does the mass of a nucleon compare with the mass of an electron?

....

c In a stable nucleus, two forces are balanced: the force that holds nuclear particles together and the force that tries to push some of them apart. Describe the two forces by completing the table.

	Name of force	Particles that the force acts on	Range of the force
Force holding particles together in nucleus			
Force pushing particles apart in nucleus			

53 Nuclear fusion

This explanation solved the mystery of the source of the Sun's energy:

For nuclear fusion to occur, two nuclei must overcome their **repulsion** and get close enough for the **attractive force** to make them join together and make a new nucleus with a larger mass. The nuclei have **kinetic energy** before and after a fusion reaction. The process of nuclear fusion releases energy, so the total kinetic energy after a reaction is greater. Fusion takes place in the core of a star because of **conditions** found there.

Explain each of the terms in bold.

repulsion: ....

....

attractive force: ....

....

kinetic energy: ....

conditions: ....

....

54 The behaviour of gases

a What four quantities are needed to fully describe the properties of a sample of a gas?

....

b The kinetic model of matter says that all matter consists of tiny particles (often molecules) in motion. The following statements explain how a gas exerts pressure on the walls of its container. Draw lines to match the two parts of each statement.

The billions of molecules in a gas . . .	. . . is related to the temperature of the gas.
The speed of the molecules . . .	. . . causes a tiny force.
As the molecules move around, they . . .	. . . move around freely in what is mostly empty space.
Each collision with the walls . . .	. . . together produce gas pressure on the walls.
The tiny forces from molecular collisions with the walls . . .	. . . collide with each other and with the walls of their container.

c Use the model of molecular collisions to explain
  i why the pressure of a gas increases when the volume of a gas is reduced, with its temperature constant

....

....

  ii why the pressure of a gas increases with temperature, when its volume stays constant

....

....

  iii why the volume of a gas increases with temperature, when its pressure stays constant

....

....

  iv what a temperature of 'absolute zero' means

....

....

55 Temperature scales

a Complete this sentence.

The absolute zero of temperature occurs at .... °C or .... K.

b Complete the table by converting each temperature from one scale to the other.

Temperature (K)	Temperature (°C)
	100
35	
	510
173	
	−15
77	

56 Stars change

Use words from the box to complete these sentences.

data	colour	H–R diagram	luminosity	white dwarf
red giants	models	main sequence	small part	

Astronomers observe stars of quite different .... and .... When stars

are plotted on a .... (a chart of luminosity against temperature), they

fall into three main groups: main sequence stars, ... stars, and .... or

supergiants. Linking .... about star populations to .... of how stars

work, astronomers conclude that stars change, and that:

An average star spends most of its lifetime as a .... star.

A star may spend a .... of its lifetime as a red giant or as

a white dwarf.

57 Main sequence stars

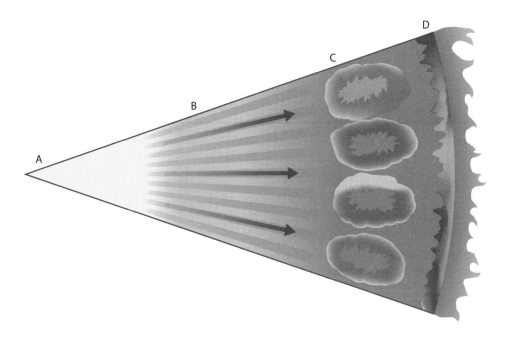

a The diagram shows the internal structure of a main sequence star.

Draw lines to match the labels and the descriptions of what happens in each part of the star.

Label	Part of the star	What happens there
A	photosphere	fusion of hydrogen takes place
B	radiative zone	energy is transferred by convection cells
C	core	energy radiates into space
D	convective zone	energy is carried by photons

b Fill in the blank to complete the sentence.

How long a star lasts in its main sequence phase depends on its .... and

....

58 Protostars

a Number the following statements in order, so that they describe the formation of a star with its solar system.

	Material further out in the disc clumps together to form planets.
	Eventually the temperature at the centre is hot enough for fusion reactions to occur and a star is born.
	A cloud of dust and hydrogen in space starts to contract, pulled together by gravity. It becomes a rotating disc.
	The temperature increases when the raw material is compressed, getting hotter and hotter at the centre.

b Use the kinetic model of matter to explain why the temperature in the centre of a protostar rises. Include in your explanation the role played by gravity.

....

....

....

....

c Explain why gravity gives any large mass of material a spherical shape.

....

....

....

59 Dying stars

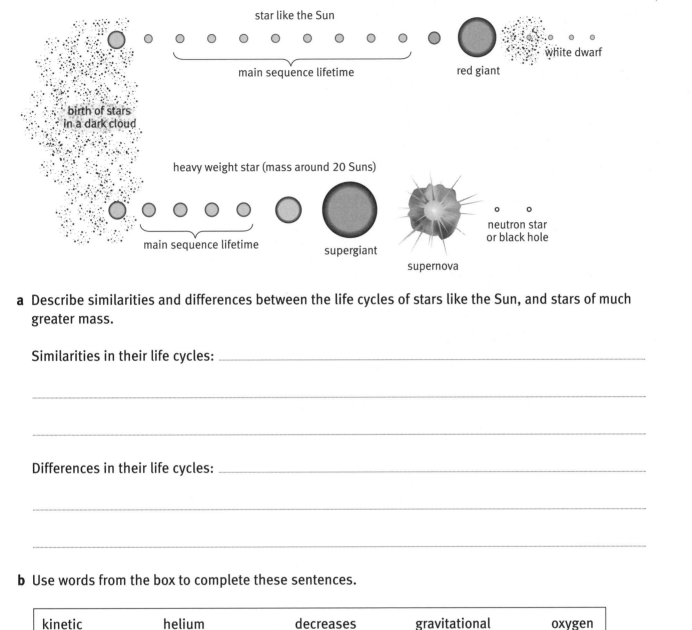

a Describe similarities and differences between the life cycles of stars like the Sun, and stars of much greater mass.

Similarities in their life cycles: ....

....

....

Differences in their life cycles: ....

....

....

b Use words from the box to complete these sentences.

kinetic	helium	decreases	gravitational	oxygen
red giant	hydrogen	carbon	increases	

As the .... of a main sequence star runs out, its core cools down and so its volume

.... The collapse transfers .... energy to ....

energy of helium nuclei, which means the star's core temperature .... This restarts the

fusion process, with .... changing to nuclei of even bigger mass such as

..........................., nitrogen, and .... The energy this releases produces a

.... or supergiant.

60 The final stages

a Fill in the blank to complete these sentences.

When a red giant runs out of helium, its mass is too small for gravity to compress its core and produce

higher temperatures, and so fusion stops. The star shrinks into a hot ....

...., which gradually cools.

b The following statements describe what happens to supergiants. Number the statements in the correct order.

	The supernova remnant becomes either a neutron star or a black hole. Remnants with the biggest masses become black holes.
	When eventually the star produces iron, it runs out of nuclear fuel. The rate of fusion in the core decreases and pressure falls.
	The weight of outer layers of the star is no longer balanced by the core's pressure. The star dramatically collapses and then explodes as a supernova.
	Fusion in a supergiant continues to produce heavier and heavier elements, because gravity causes such high pressures in the core of massive stars.

61 Astronomy today

a Describe two ways that astronomers work with local or remote telescopes.

1 ....

....

....

2 ....

....

....

b What advantages does computer control of telescopes offer?

_____

_____

_____

_____

c Read the following statements about using telescopes outside the Earth's atmosphere. Put an A next to those that are advantages, and a D next to those that are disadvantages.

Advantage or disadvantage?	Statement
	They avoid absorption and refraction effects of the atmosphere.
	They are expensive to build and launch.
	They can detect radiation from astronomical objects in parts of the electromagnetic spectrum that are strongly absorbed by the atmosphere.
	Servicing relies on space programmes that astronomers cannot control.
	Orbit allows imaging from all parts of the sky.
	If things go wrong it is much harder to repair them.
	Instruments can quickly become out of date and are not easily replaced.
	Launching limits the size of the telescope.

d Give three reasons why international collaboration is common in astronomy.

1 _____

2 _____

3 _____

62 Choosing a site

a Describe two factors that affect the quality of 'seeing' that influence the choice of site for astronomical observatories.

1 ....

....

2 ....

....

b List four factors of other types that are important in building and operating an observatory.

1 ....

2 ....

3 ....

4 ....